A  scles

ONE WEEK LOAN

SEVENTH EDITION

Atlas of Skeletal Muscles

Robert J. Stone
Suffolk County Community College

Judith A. Stone
Suffolk County Community College

The McGraw·Hill Companies

Mc
Graw
Hill
) Connect
Learn
Succeed™

ATLAS OF SKELETAL MUSCLES, SEVENTH EDITION

Published by McGraw-Hill, a business unit of The McGraw-Hill Companies, Inc., 1221 Avenue of the Americas, New York, NY 10020. Copyright © 2012 by The McGraw-Hill Companies, Inc. All rights reserved. Previous editions © 2009, 2006 and 2003. No part of this publication may be reproduced or distributed in any form or by any means, or stored in a database or retrieval system, without the prior written consent of The McGraw-Hill Companies, Inc., including, but not limited to, in any network or other electronic storage or transmission, or broadcast for distance learning.

Some ancillaries, including electronic and print components, may not be available to customers outside the United States.

This book is printed on acid-free paper.

1 2 3 4 5 6 7 8 9 0 QDB/QDB 1 0 9 8 7 6 5 4 3 2 1

ISBN 978-0-07-131668-2
MHID 0-07-131668-X

www.mhhe.com

Dedication

To Karen, Andrew, and Laura.

Contents

CHAPTER FOUR

Muscles of the Neck 61

CHAPTER FIVE

Muscles of the Trunk 83

CHAPTER SIX

Muscles of the Shoulder and Arm 109

CHAPTER SEVEN
Muscles of the Forearm and Hand 131

CHAPTER EIGHT
Muscles of the Hip and Thigh 163

Preface

This book is a study guide and reference for the anatomy and actions of human skeletal muscles. It is designed for use by students of anatomy, physical education, and health-related fields. It also serves as a compact reference for the practicing professional.

The first chapter presents photographic illustrations of the major features of the skeleton. These photos have been selectively enhanced to emphasize important features. They are thus a hybrid between drawings and photographs. A new feature in this edition is the addition of joints with their corresponding tendons and ligaments drawn on bone photographs.

The second chapter describes through illustration and description the various movements of the body.

In chapters 3 through 9, the origin, insertion, action, and innervation of the skeletal muscles are described, and each muscle is presented on a separate page. The spinal cord levels of the nerve fibers that innervate each muscle are included in parentheses after the name of each nerve.

The drawings include the following important features:

1. Muscle fibers are shown in red.
2. Tendons and aponeuroses are shown.
3. Where muscle fibers attach on the undersurface of bone and cartilage, the latter have been rendered semi-transparent.
4. Adjacent structures are shown for easy visualization of muscle location.
5. The muscle is shown in red in boxed inserts in relationship to other muscles.
6. Boxed inserts present on many pages illustrate anatomical and/or functional relationships of muscle groups or enlargements for detail.

These features aid in visual orientation and understanding of the action of the muscles. Notes have been included on many pages to show how muscles are used. Relationships among many of the muscles have also been indicated where appropriate.

Some users of previous editions have advised that some of the smaller muscles should be enlarged and shown with less skeletal background. We have purposely standardized the skeletal views to allow an appreciation of the relative sizes and positions of the muscles. Skeletal muscles, at the gross level, are relatively simple anatomical structures, so very little additional information would be included by enlargement, and many comparative relationships would be lost.

Our primary objective is to describe the muscles moving the skeleton, therefore we have not included the muscles of the peritoneum, eye, tympanic cavity, tongue, larynx, pharynx, or palate.

We fondly remember the late Mr. George Boykin, for many years the jolly proprietor the gross anatomy laboratories at the State University of New York at Stony Brook. His help and encouragement were instrumental in the early stages of this project. We also thank Dr. Jagdish Gidwani, Mr. Vincent Verdisco and Dr. Peter Smith for their technical advice and the many students who have offered valuable suggestions over the years.

We would also like to thank the many reviewers who have made helpful suggestions for improving past editions of this atlas, as well as Joseph Myers, University of North Carolina at Chapel Hill, Curt Walker, Dixie State College, and Brent Branstetter, Purdue University for their input on the seventh edition.

Robert J. Stone

Judith A. Stone

CHAPTER ONE

The Skeleton

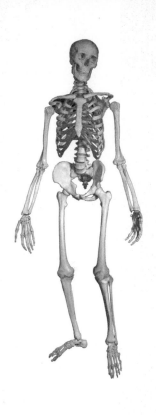

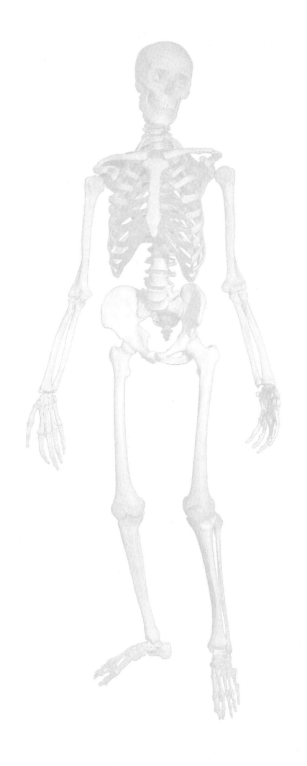

SKULL—ANTERIOR VIEW

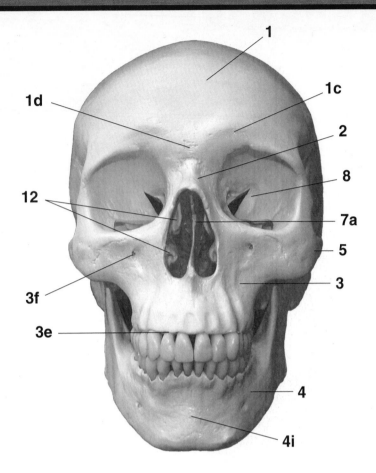

1. Frontal bone
1c. Superciliary arch
1d. Glabella
2. Nasal bone
3e–h. Maxilla
3e. Alveolar border
3f. Infraorbital foramen
3g. Incisive foramen
3h. Palatine process
4. Mandible
4i. Symphysis
5. Zygomatic bone
7a. Perpendicular plate of ethmoid bone
8. Sphenoid
8b. Lateral pterygoid plate

8d. Medial pterygoid plate
8e. Foramen ovale
9b–f. Temporal bone
9b. Mastoid process (temporal bone)
9c. Foramen lacerum
9e. Zygomatic process (temporal bone)
9f. Carotid canal
11a. Superior nuchal line (occipital bone)
11b. Inferior nuchal line (occipital bone)
11c. External occipital protuberance (occipital bone)
11d. Occipital condyle (occipital bone)
11e. External occipital crest (occipital bone)
11f. Jugular foramen
11g. Foramen magnum
12. Turbinates
13. Palatine bone
14. Vomer

SKULL—INFERIOR (BASAL) VIEW

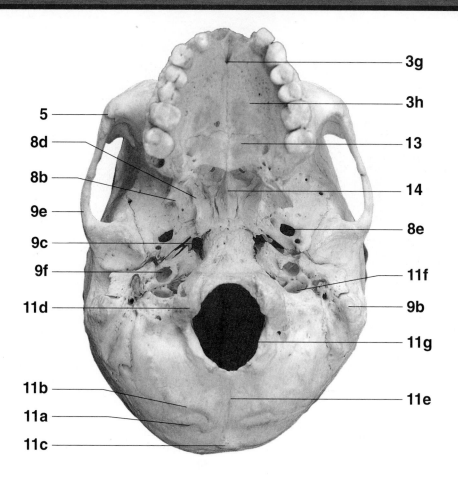

3g

3h

5

8d

8b

9e

9c

9f

11d

11b

11a

11c

13

14

8e

11f

9b

11g

11e

SKULL—LATERAL VIEW

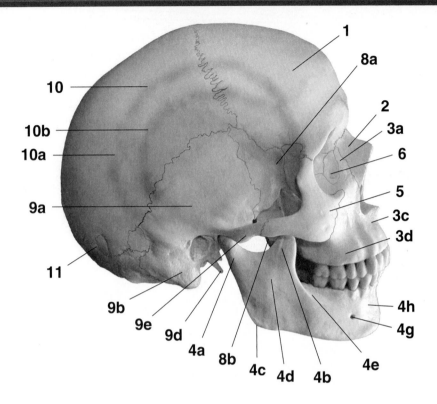

1. Frontal bone
2. Nasal bone
3a–d. Maxilla
  3a. Frontal process
  3c. Incisive process
  3d. Canine fossa
4a–h. Mandible
  4a. Neck of condyle
  4b. Coronoid process
  4c. Angle
  4d. Ramus
  4e. Oblique line
  4g. Mental foramen
  4h. Incisive fossa
5. Zygomatic bone
6. Lacrimal bone
8a. Greater wing of sphenoid bone
8b. Lateral pterygoid plate of sphenoid bone

9a–e. Temporal bone
  9a. Temporal fossa (squamous part of temporal bone)
  9b. Mastoid process
  9d. Styloid process
  9e. Zygomatic process
10. Parietal bone
10a. Superior temporal line
10b. Inferior temporal line
11. Occipital bone

Note: The zygomatic arch is formed by the zygomatic process of the temporal bone meeting the zygomatic bone.

VERTEBRAL COLUMN—LATERAL VIEW

C1–7. Cervical vertebrae
C1. Atlas
C2. Axis
C7. Seventh cervical vertebra
T1–12. Thoracic vertebrae
T1. First thoracic vertebra
T12. Twelfth thoracic vertebra
L1–5. Lumbar vertebrae
L1. First lumbar vertebra
L5. Fifth lumbar vertebra
S. Sacrum
Co. Coccyx

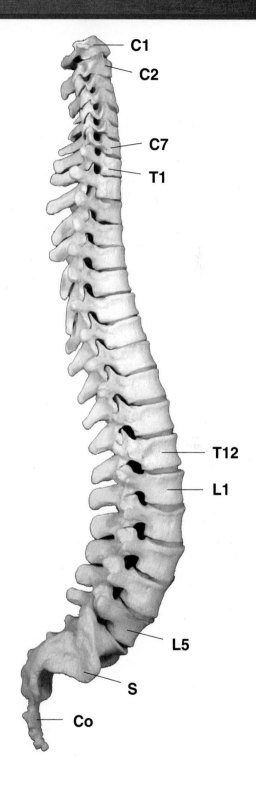

ATLAS AND AXIS

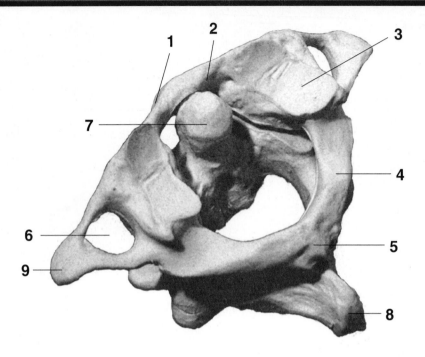

Atlas

1. Anterior tubercle
2. Anterior arch
3. Superior articular facet
4. Posterior arch
5. Posterior tubercle
6. Transverse foramen
7. Transverse process

Axis

8. Dens (odontoid process)
9. Spinous process

THORACIC VERTEBRAE—LATERAL VIEW

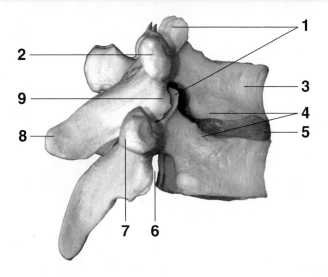

1. Superior articular process
2. Transverse process
3. Body
4. Demifacets (for ribs)
5. Disk
6. Inferior vertebral notch
7. Facet (for rib tubercle)
8. Spinous process
9. Inferior articular process

LUMBAR VERTEBRA—SUPERIOR VIEW

1. Spinous process
2. Mammillary process
3. Transverse process
4. Lamina
5. Vertebral foramen
6. Pedicle
7. Body (centrum)

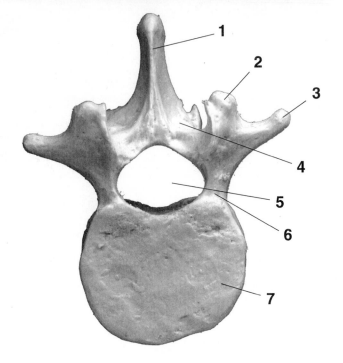

SACRUM—POSTERIOR VIEW

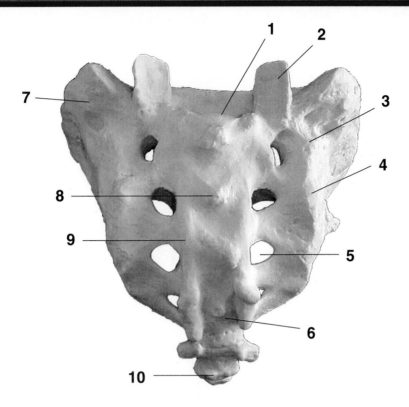

1. Sacral canal
2. Superior articular facet
3. Sacral tuberosity
4. Lateral crest
5. Posterior sacral foramen
6. Sacral hiatus (opening of sacral canal)
7. Auricular surface
8. Median crest
9. Intermediate crest
10. Coccyx

SACRUM—PELVIC VIEW

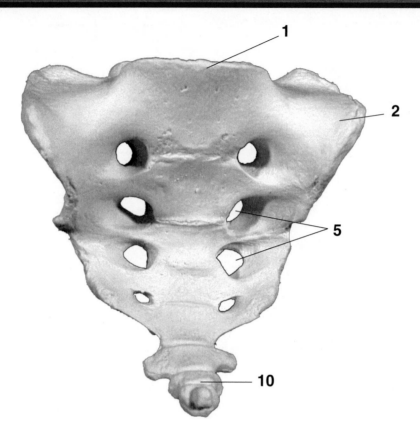

1. Promontory
2. Ala
5. Sacral foramina
10. Coccyx

STERNUM AND CLAVICLE WITH SCAPULA

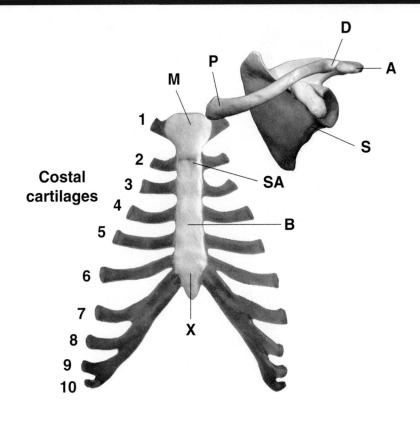

Clavicle
P = proximal end (sternal)
D = distal end (acromial or lateral)
Scapula (S)
A = acromion
Sternum
M = manubrium
SA = sternal angle
B = body
X = xiphoid process

The proximal (sternal) end of the clavicle forms the sternoclavicular joint with the manubrium of the sternum.

The distal (scapular) end forms the acromioclavicular joint with the acromion of the scapula. This is the only bony articulation of the upper limb with the torso.

The second rib cartilage articulates at the sternal angle between the manubrium and the body of the sternum.

The cartilages of ribs 7–10 are fused to form the costal arch.

HYOID BONE—SUPERIOR VIEW

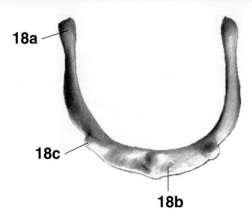

18a–c. Hyoid bone
  18a. Greater horn
  18b. Body
  18c. Lesser horn

The greater horn of the hyoid bone attaches to the styloid processes of the temporal bones by the styloid ligaments. It attaches to the thyroid cartilage by the thyrohyoid ligament and supports the upper respiratory tract.

RIB ARTICULATIONS

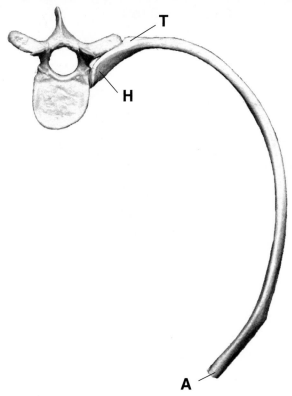

The head (H) of a rib articulates with the demifacets of two adjacent vertebrae. The tubercle (T) joins the facet of the transverse process of the upper vertebra. The anterior end (A) meets the costal cartilage, which then joins the sternum.

SCAPULA AND HUMERUS—ANTERIOR VIEW

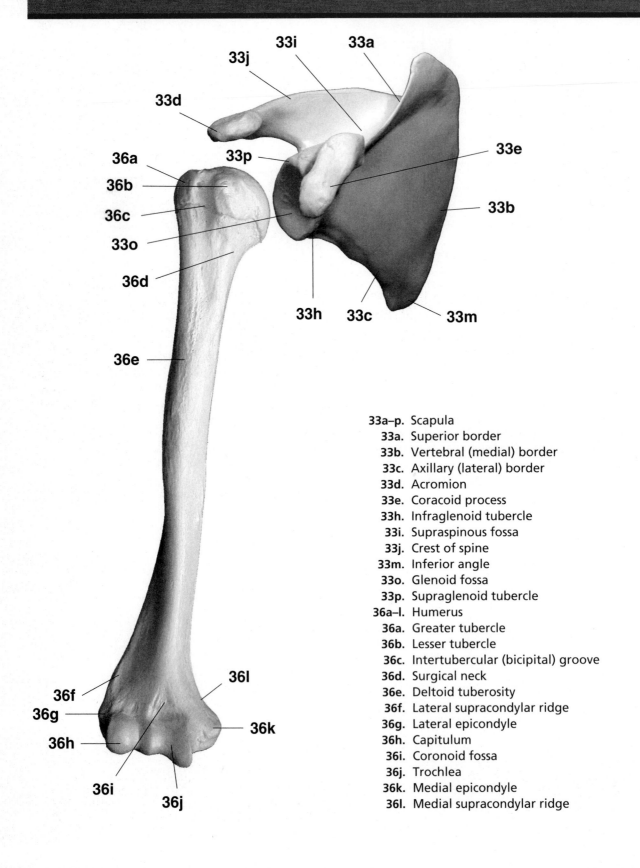

33a–p. Scapula
  33a. Superior border
  33b. Vertebral (medial) border
  33c. Axillary (lateral) border
  33d. Acromion
  33e. Coracoid process
  33h. Infraglenoid tubercle
  33i. Supraspinous fossa
  33j. Crest of spine
  33m. Inferior angle
  33o. Glenoid fossa
  33p. Supraglenoid tubercle
36a–l. Humerus
  36a. Greater tubercle
  36b. Lesser tubercle
  36c. Intertubercular (bicipital) groove
  36d. Surgical neck
  36e. Deltoid tuberosity
  36f. Lateral supracondylar ridge
  36g. Lateral epicondyle
  36h. Capitulum
  36i. Coronoid fossa
  36j. Trochlea
  36k. Medial epicondyle
  36l. Medial supracondylar ridge

SCAPULA AND HUMERUS—POSTERIOR VIEW

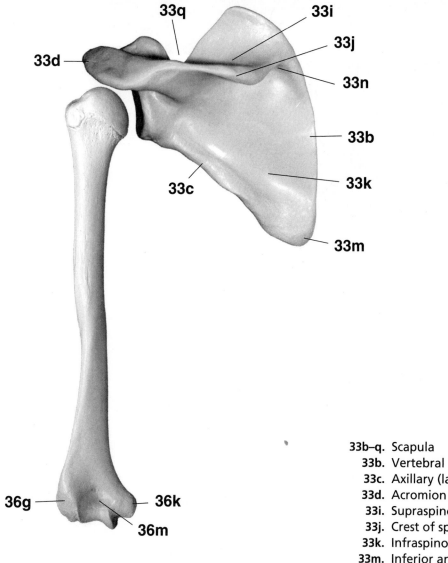

33q
33i
33j
33d
33n
33b
33c
33k
33m
36g
36k
36m

33b–q. Scapula
  33b. Vertebral (medial) border
  33c. Axillary (lateral) border
  33d. Acromion
  33i. Supraspinous fossa
  33j. Crest of spine
  33k. Infraspinous fossa
  33m. Inferior angle
  33n. Root of spine
  33q. Suprascapular notch
36g–m. Humerus
  36g. Lateral epicondyle
  36k. Medial epicondyle
  36m. Olecranon fossa

SHOULDER ANTERIOR WITH LIGAMENTS—ANTERIOR VIEW

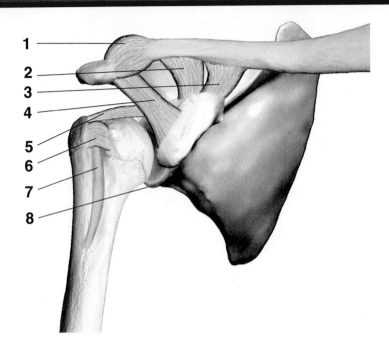

1. Acromioclavicular capsule and ligament
2. Trapezoid part of coracoclavicular ligament
3. Conoid part of coracoclavicular ligament
4. Coracoacromial ligament
5. Coracohumeral ligament
6. Transverse humeral ligament
7. Biceps brachii tendon (long head)
8. Glenoid labrum

ELBOW—ANTERIOR VIEW

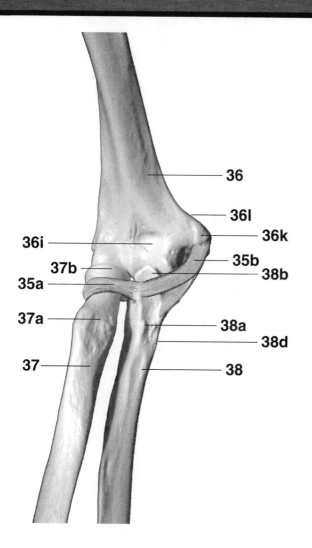

35a. Annular ligament
35b. Ulnar (medial) collateral ligament
36i–l. Humerus
36i. Coronoid fossa
36k. Medial epicondyle
36l. Medial supracondylar ridge
37a, b. Radius
37a. Radial tuberosity
37b. Head of radius
38a–d. Ulna
38a. Ulnar tuberosity
38b. Coronoid process
38d. Supinator crest

FOREARM—POSTERIOR VIEW

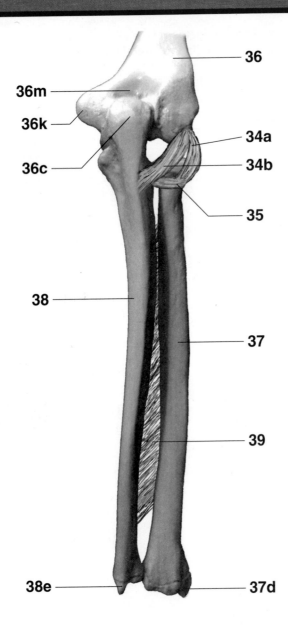

34a. Radial (lateral) collateral ligament
34b. Ulnar (lateral) collateral ligament
35. Annular ligament
36k–m. Humerus
36k. Medial epicondyle
36m. Olecranon fossa
37. Radius
37d. Styloid process of radius
38c–e. Ulna
38c. Olecranon
38e. Styloid process of ulna
39. Interosseous membrane

HAND—PALMAR VIEW

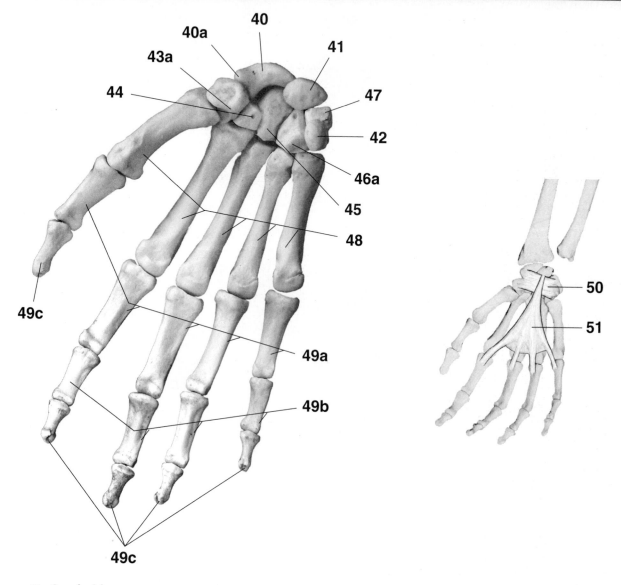

40. Scaphoid
40a. Tubercle of scaphoid
41. Lunate
42. Pisiform
43a. Tubercle of trapezium
44. Trapezoid
45. Capitate
46a. Hook of hamate
47. Triquetrum
48. Metacarpals

49a. Proximal (first) phalanges
49b. Middle (second) phalanges
49c. Distal (third) phalanges
50. Flexor retinaculum
51. Palmar aponeurosis

HAND—DORSAL VIEW

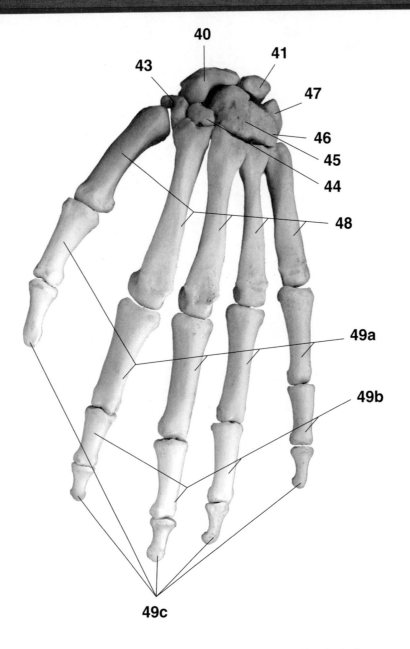

40. Scaphoid
41. Lunate
43. Trapezium
44. Trapezoid
45. Capitate
46. Hamate
47. Triquetrum
48. Metacarpals

49a. Proximal (first) phalanges
49b. Middle (second) phalanges
49c. Distal (third) phalanges

LIGAMENTS OF THE WRIST AND HAND—PALMAR VIEW

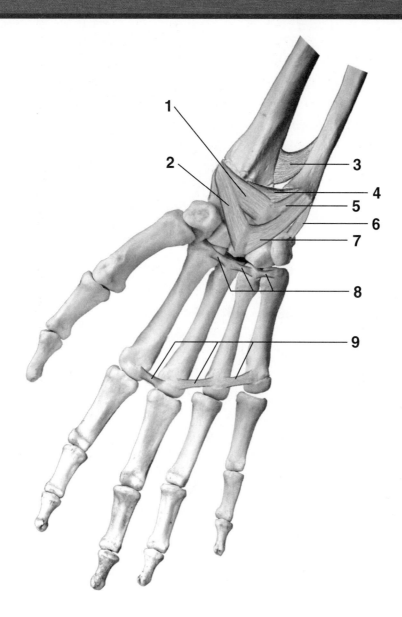

1. Long radiolunate
2. Palmar radiocarpal
3. Interosseous membrane
4. Joint capsule
5. Ulnolunate
6. Ulnotriquetral
7. Palmar ulnocarpal ligaments
8. Palmar metacarpal ligaments
9. Deep transverse metacarpal ligaments

PELVIS—ANTERIOR VIEW

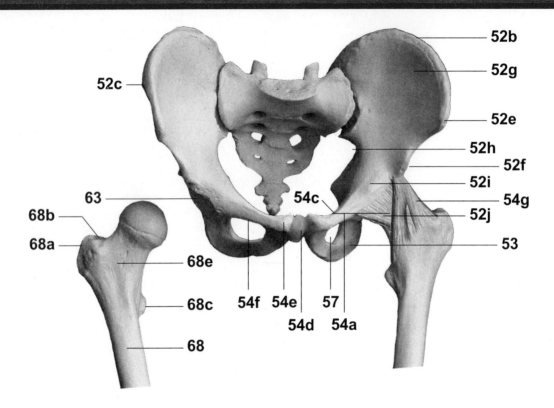

52b–j. Ilium
  52b. Iliac crest
  52c. Iliac tubercle
  52e. Anterior superior iliac spine
  52f. Anterior inferior iliac spine
  52g. Iliac fossa
  52h. Arcuate line
  52i. Iliopectineal eminence
  52j. Iliofemoral ligament (Y ligament of Bigelow)
  53. Ischium
54a–e. Pubis
  54a. Superior ramus
  54b. Inferior ramus
  54c. Pubic crest
  54d. Pubic symphysis
  54e. Pubic tubercle
  54f. Obturator crest
  54g. Pubofemoral ligament
  57. Obturator foramen
  63. Acetabulum
68a–e. Femur
  68a. Greater trochanter
  68b. Trochanteric fossa
  68c. Lesser trochanter
  68e. Intertrochanteric line

PELVIS—THREE QUARTER POSTERIOR VIEW

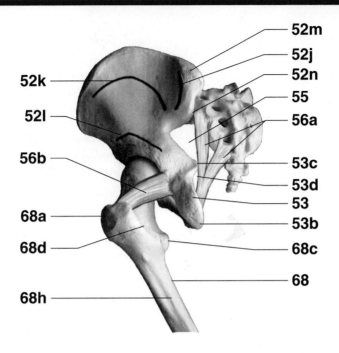

52k
52l
56b
68a
68d
68h

52m
52j
52n
55
56a
53c
53d
53
53b
68c
68

52j–n. Ilium
  52j. Posterior gluteal line
  52k. Middle gluteal line
  52l. Inferior gluteal line
  52m. Posterior superior iliac spine
  52n. Posterior inferior iliac spine
53b–d. Ischium
  53b. Ischial tuberosity
  53c. Ischial spine
  53d. Lesser sciatic notch (lesser sciatic foramen)
  55. Greater sciatic notch (greater sciatic foramen)
  56a. Sacrotuberous ligament
  56b. Ischiofemoral ligament
68a–h. Femur
  68a. Greater trochanter
  68c. Lesser trochanter
  68d. Intertrochanteric crest
  68g. Gluteal tuberosity
  68h. Linea aspera

*Note: These bony notches are converted to foramina by the two ligaments.

FEMUR—POSTERIOR VIEW

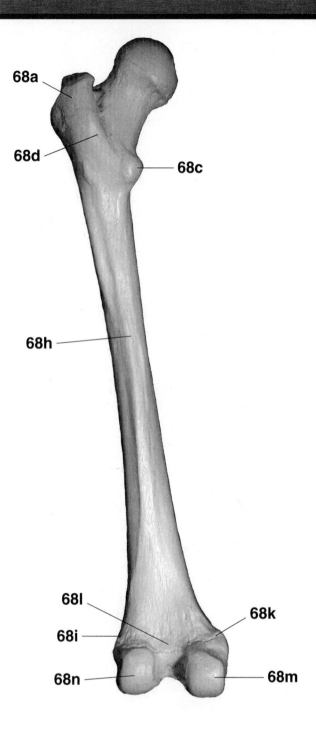

68a–n. Femur
68a. Greater trochanter
68c. Lesser trochanter
68d. Intertrochanteric crest
68h. Linea aspera
68i. Lateral supracondylar ridge
68k. Adductor tubercle
68l. Popliteal surface
68m. Medial condyle
68n. Lateral condyle

ACETABULAR LIGAMENTS

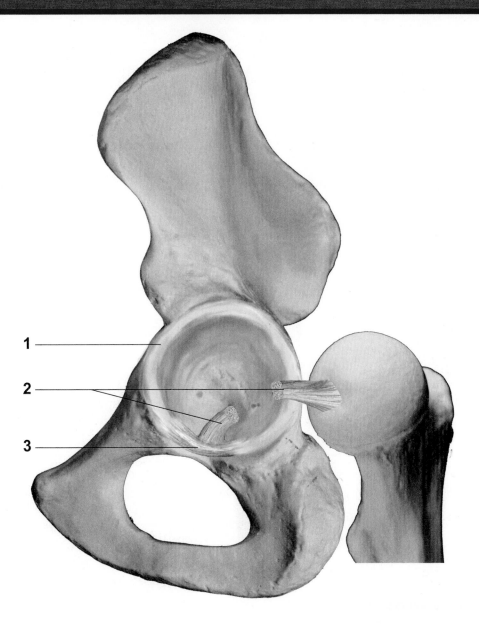

1. Acetabular labrum
2. Ligamentum teres
3. Transverse acetabular ligament

KNEE—ANTERIOR VIEW

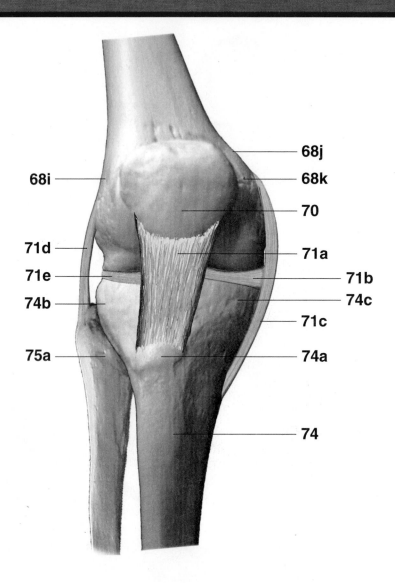

68i – k. Femur
   68i. Lateral supracondylar ridge
   68j. Medial supracondylar ridge
   68k. Adductor tubercle
   70. Patella
   71a. Patellar ligament
   71b. Medial meniscus
   71c. Medial collateral ligament
   71d. Lateral collateral ligament
74a – c. Tibia
   74a. Tibial tuberosity
   74b. Lateral condyle
   74c. Medial condyle
   75a. Head of fibula

Note: The quadriceps tendon continues through the patella and becomes the patellar ligament.

KNEE CRUCIATE MENISCI—INTERIOR VIEW

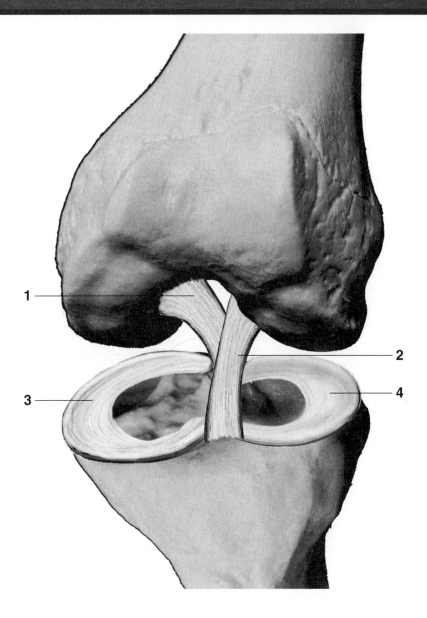

1. Posterior cruciate ligament
2. Anterior cruciate ligament
3. Medial meniscus
4. Lateral meniscus

ANKLE AND FOOT—ANTEROLATERAL VIEW

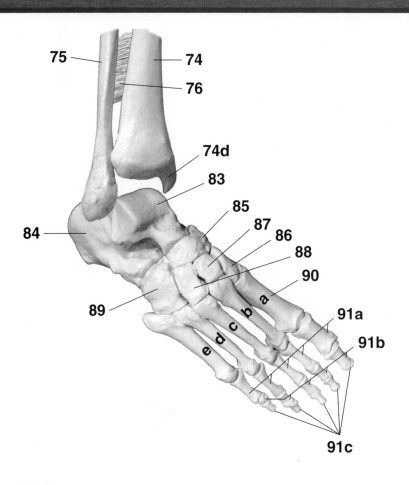

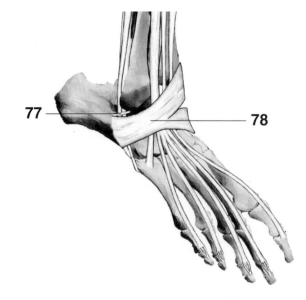

- **74.** Tibia
- **74d.** Medial malleolus of tibia
- **75.** Fibula
- **76.** Interosseous membrane
- **77.** Lateral talocalcaneal ligament
- **78.** Inferior extensor retinaculum
- **83.** Talus
- **84.** Calcaneus
- **85.** Navicular
- **86.** Medial cuneiform
- **87.** Intermediate cuneiform
- **88.** Lateral cuneiform
- **89.** Cuboid
- **90a–e.** Metatarsal bones
- **90a.** First metatarsal
- **90b.** Second metatarsal
- **90c.** Third metatarsal
- **90d.** Fourth metatarsal
- **90e.** Fifth metatarsal
- **91a.** Proximal phalanges
- **91b.** Middle phalanges
- **91c.** Distal phalanges

FOOT—PLANTAR VIEW

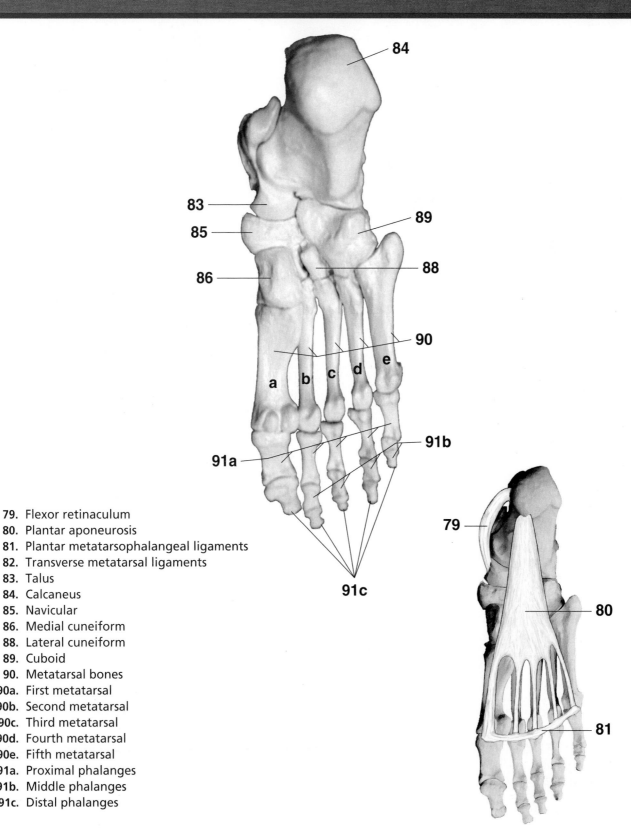

79. Flexor retinaculum
80. Plantar aponeurosis
81. Plantar metatarsophalangeal ligaments
82. Transverse metatarsal ligaments
83. Talus
84. Calcaneus
85. Navicular
86. Medial cuneiform
88. Lateral cuneiform
89. Cuboid
90. Metatarsal bones
90a. First metatarsal
90b. Second metatarsal
90c. Third metatarsal
90d. Fourth metatarsal
90e. Fifth metatarsal
91a. Proximal phalanges
91b. Middle phalanges
91c. Distal phalanges

ANKLE—MEDIAL AND LATERAL VIEWS

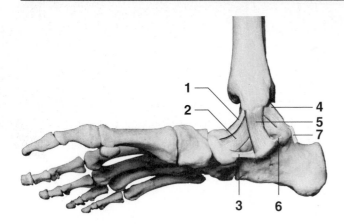

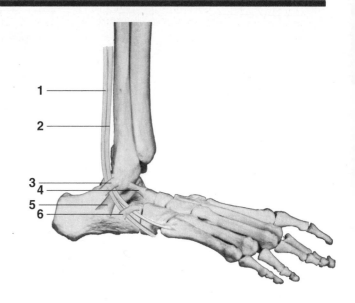

Medial View

1. Anterior tibiotalar (deltoid) ligament
2. Tibionavicular (deltoid) ligament
3. Plantar calcaneonavicular ligament
4. Posterior tibiotalar (deltoid) ligament
5. Tibiocalcaneal ligament
6. Medial talocalcaneal ligament
7. Posterior talocalcaneal ligament

Lateral View

1. Fibularis (peroneus) longus tendon
2. Fibularis (peroneus) brevis tendon
3. Superior fibular (peroneal) retinaculum
4. Anterior talofibular ligament
5. Calcaneofibular ligament
6. Inferior fibular (peroneal) retinaculum

CHAPTER TWO

Movements of the Body

Anatomical position—A subject in the anatomical position is standing erect with the head, eyes, and toes facing forward and the arms hanging straight at the sides with the palms of the hands facing forward.

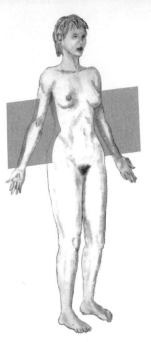

F I G U R E 2.2
Coronal (frontal) planes—Pass vertically through the body from side to side. They divide the body from front to back.

F I G U R E 2.1
Median or midsagittal plane—Passes vertically through the body from anterior (front) to posterior (back). It divides the body into right and left sides. Other sagittal planes are parallel to this plane.

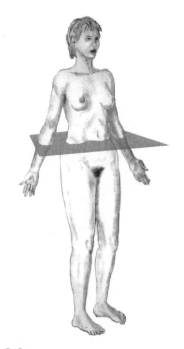

F I G U R E 2.3
Transverse planes (cross sections)—Pass horizontally through the body parallel to the ground.

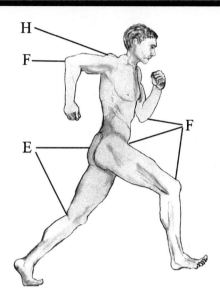

FIGURE 2.4
Flexion-extension—Starting from the anatomical position, movement is anterior or posterior in a sagittal plane. In the hinge joints *flexion* results in a decrease in the angle made by the bones in the joint, and *extension* brings the bones toward a 180° angle. Hyperextension is permitted in the shoulder, wrist, and vertebrae where movement continues posterior to anatomical position. In the knee, hip, and elbow, hyperextension is prevented by bone structure or ligaments.

FIGURE 2.5
Lateral flexion—The torso (or head) bends laterally in the coronal plane.

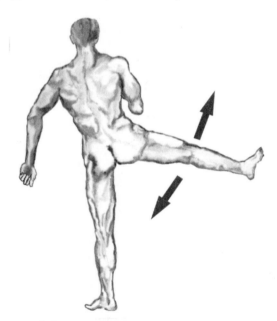

FIGURE 2.6
Abduction—The right leg moves laterally in the coronal plane.
Adduction—The leg is returned medially in the coronal plane.

Note: If the foot is fixed, this motion results in tilting the pelvis upward on the opposite side.

FIGURE 2.7
Medial rotation—The anterior of the arm (or thigh) is moved toward the median plane.
Lateral rotation—The anterior of the arm (or thigh) is moved away from the median plane.

MOVEMENTS OF THE SCAPULA

FIGURE 2.8
Elevation—The right scapula of this figure is drawn superiorly against the resistance of the rock.

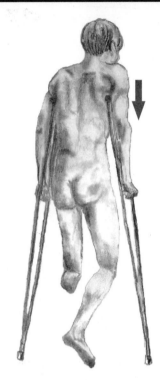

FIGURE 2.9
Depression—The right scapula of this figure is pushing the arm inferiorly.

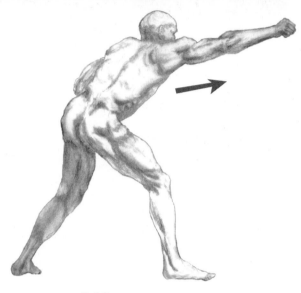

FIGURE 2.10
Protraction—The scapula pushes the arm forward in a sagittal plane.

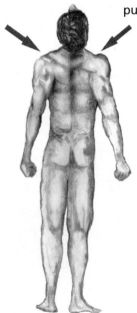

FIGURE 2.11
Retraction—The scapula is pulled back from protraction in a sagittal plane. The scapula slides around the ribs toward the median plane, so it becomes adduction.

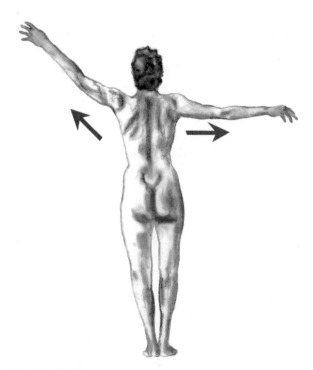

FIGURE 2.12
Rotation—For abduction of the arm to continue above the height of the shoulder, the scapula must rotate on its axis so that the glenoid fossa turns upward.

MOVEMENTS OF THE HAND AND FOREARM

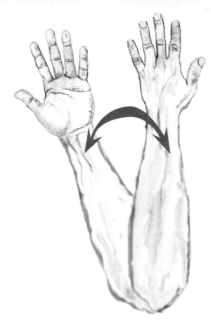

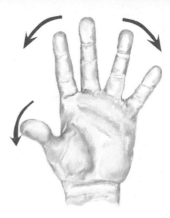

FIGURE 2.14
Abduction—The fingers are moved away from the midline of the hand.

FIGURE 2.13
Pronation—The forearm is rotated away from the anatomical position so that the palm turns medially, then posteriorly. If the forearm is flexed at the elbow, then the palm turns inferiorly.
Supination—The forearm is rotated so that the palm turns anteriorly (or superiorly if the forearm is flexed). Also see figure 2.19.

FIGURE 2.16
Adduction—The fingers are moved toward the midline of the hand.

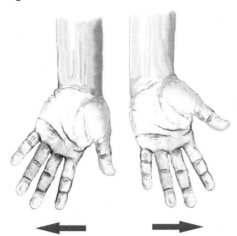

FIGURE 2.15
Radial deviation (abduction)—The hand, at the wrist, is moved laterally toward the radius. In anatomical position this moves the hand away from the body in the coronal plane.
Ulnar deviation (adduction)—The hand, at the wrist, is moved medially toward the ulna.

FIGURE 2.17
Opposition—The saddle joint between the trapezium and first metacarpal allows the thumb to turn so its anterior surface can touch the anterior surfaces of the four fingers when they are partially flexed.

MOVEMENTS OF THE FOOT

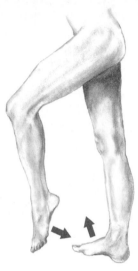

FIGURE 2.18
Dorsiflexion—Elevating the foot, decreasing the angle between the foot and the leg.
Plantar flexion—Depressing the foot, increasing the angle between the foot and the leg.

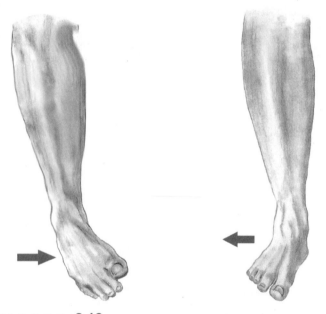

FIGURE 2.19
Inversion-eversion—During *inversion* the sole of the foot turns medially, and during *eversion* it returns or turns slightly laterally. Most of this movement is permitted by the gliding of the intertarsal and tarsometatarsal joints but because the articulating surface of the talus is narrower posteriorly, it permits some lateral movement when the ankle is plantar flexed.

CHAPTER THREE

Muscles of the Face and Head

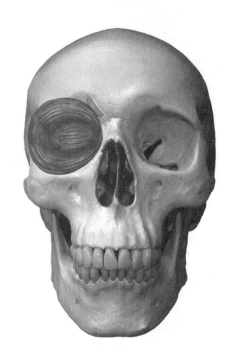

OCCIPITOFRONTALIS*

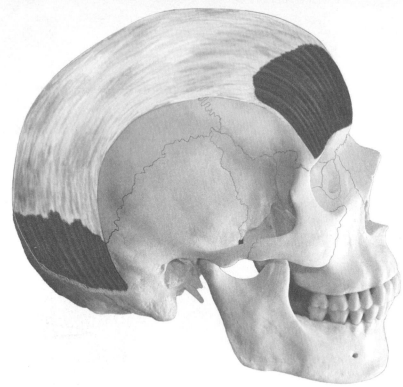

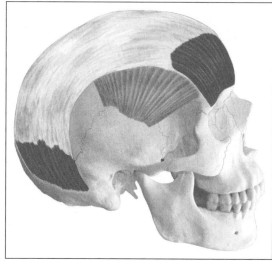

Epicranius

Skull—lateral view

Occipital belly *(occipitalis)*

■ **Origin**
Lateral two-thirds of superior nuchal line of occipital bone, mastoid process of temporal bone

■ **Insertion**
Galea aponeurotica (an intermediate tendon leading to frontal belly)

■ **Action**
Draws back scalp, aids frontal belly to wrinkle forehead and raise eyebrows

■ **Nerve**
Posterior auricular branch of facial nerve

Frontal belly *(frontalis)*

■ **Origin**
Galea aponeurotica

■ **Insertion**
Fascia of facial muscles and skin above nose and eyes

■ **Action**
Draws back scalp, wrinkles forehead, raises eyebrows

■ **Nerve**
Temporal branches of facial nerve

*The occipitofrontalis and temperoparietalis are collectively called epicraneus.

TEMPOROPARIETALIS

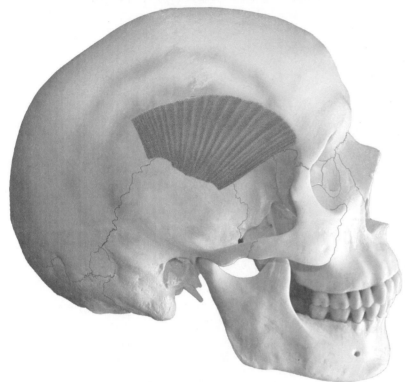

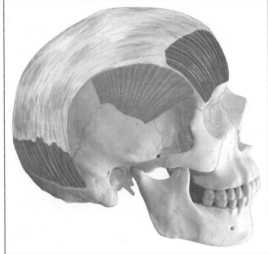

Epicranius

Skull—lateral view

■ **Origin**
Fascia over ear

■ **Insertion**
Lateral border of galea aponeurotica

■ **Action**
Raises ears, tightens scalp

■ **Nerve**
Temporal branch of facial nerve

*The occipitofrontalis and temperoparietalis are collectively called epicraneus.

AURICULARIS ANTERIOR, SUPERIOR, POSTERIOR

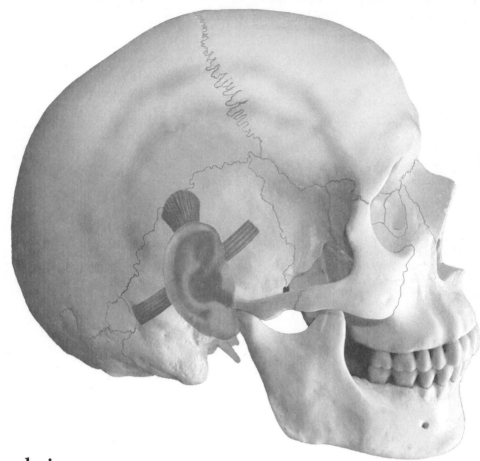

Skull—lateral view

Auricularis anterior

■ **Origin**
Fascia in temporal region

■ **Insertion**
Anterior to helix of ear

■ **Action**
Draws ear forward in some individuals, moves scalp*

■ **Nerve**
Temporal branch of facial nerve

Auricularis superior

■ **Origin**
Fascia in temporal region

■ **Insertion**
Superior part of ear

*This muscle is nonfunctional in most people.

■ **Action**
Draws ear upward in some individuals, moves scalp*

■ **Nerve**
Temporal branch of facial nerve

Auricularis posterior

■ **Origin**
Mastoid area of temporal bone

■ **Insertion**
Posterior part of ear

■ **Action**
Draws ear upward in some individuals*

■ **Nerve**
Posterior auricular branch of facial nerve

ORBICULARIS OCULI

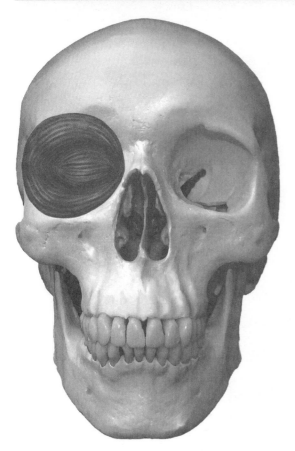

Lacrimal part—lateral view

Orbital and palpebral parts—anterior view

Skull—anterior view

Orbital part

■ Origin
Frontal bone, maxilla (medial margin of orbit)

■ Insertion
Continues around orbit and returns to origin

■ Action
Strong closure of eyelids

■ Nerve
Temporal and zygomatic branches of facial nerve

Palpebral part *(in eyelids)*

■ Origin
Medial palpebral ligament

■ Insertion
Lateral palpebral ligament into zygomatic bone

■ Action
Gentle closure of eyelids

■ Nerve
Temporal and zygomatic branches of facial nerve

Lacrimal part *(behind medial palpebral ligament and lacrimal sac)*

■ Origin
Lacrimal bone

■ Insertion
Lateral palpebral raphe

■ Action
Draws lacrimal canals onto surface of eye

■ Nerve
Temporal and zygomatic branches of facial nerve

LEVATOR PALPEBRAE SUPERIORIS

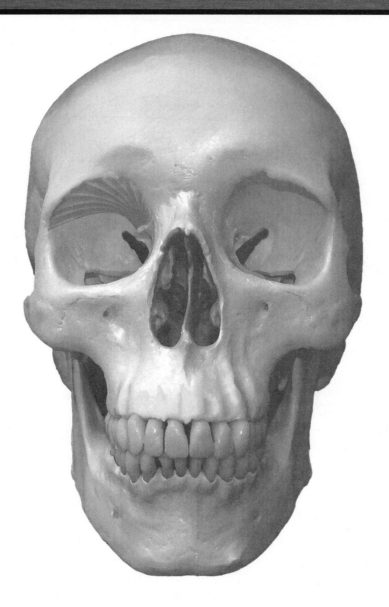

Skull—anterior view

■ **Origin**
Inferior surface of lesser wing of sphenoid

■ **Insertion**
Skin of upper eyelid

■ **Action**
Raises upper eyelid

■ **Nerve**
Oculomotor nerve

CORRUGATOR SUPERCILII

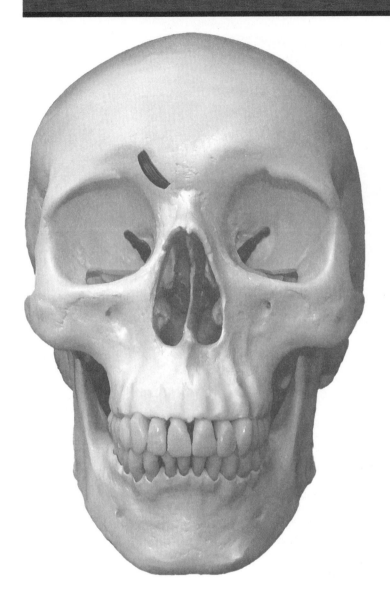

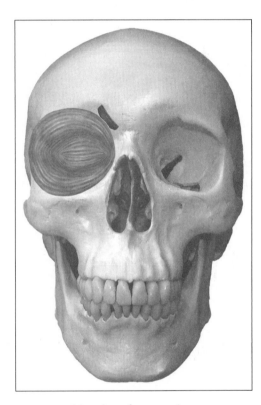

Muscles of eye region

Skull—anterior view

■ Origin
Medial end of superciliary arch

■ Insertion
Deep surface of skin under medial portion of eyebrows

■ Action
Draws eyebrows downward and medially

■ Nerve
Temporal branch of facial nerve

PROCERUS

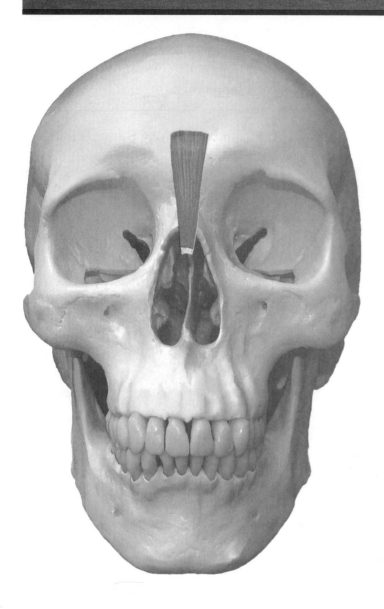

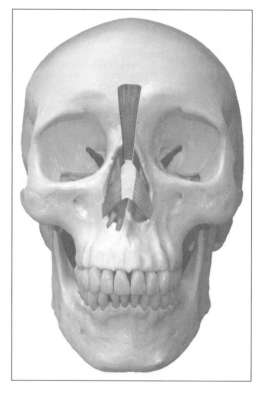

Muscles of nose

Skull—anterior view

■ **Origin**
Fascia over nasal bone and lateral nasal cartilage

■ **Insertion**
Skin between eyebrows

■ **Action**
Draws down medial part of eyebrows, wrinkles nose

■ **Nerve**
Buccal branches of facial nerve

NASALIS

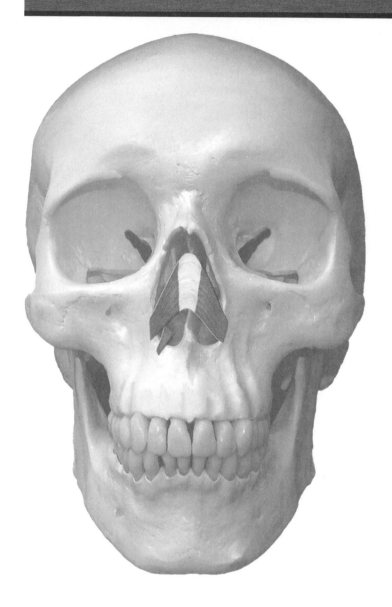

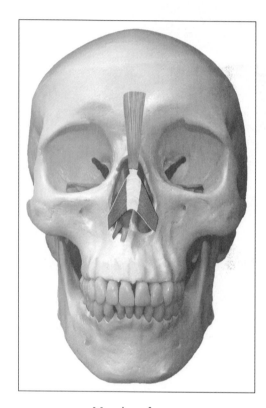

Muscles of nose

Skull—anterior view

Transverse part

■ **Origin**
Middle of maxilla

■ **Insertion**
Muscle of opposite side over bridge of nose

Alar part

■ **Origin**
Greater alar cartilage, skin on nose

■ **Insertion**
Skin at point of nose

■ **Action**
Both parts maintain opening of external nares during forceful inspiration

■ **Nerve**
Buccal branches of facial nerve

DEPRESSOR SEPTI

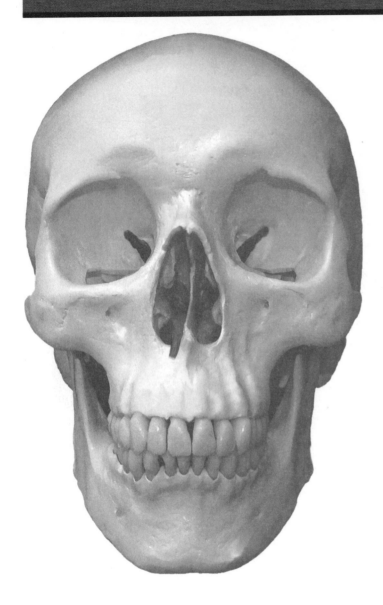

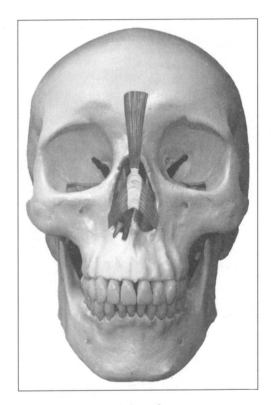

Muscles of nose

Skull—anterior view

■ **Origin**
Incisive fossa of maxilla

■ **Insertion**
Nasal septum and ala

■ **Action**
Constricts nares

■ **Nerve**
Buccal branches of facial nerve

ORBICULARIS ORIS

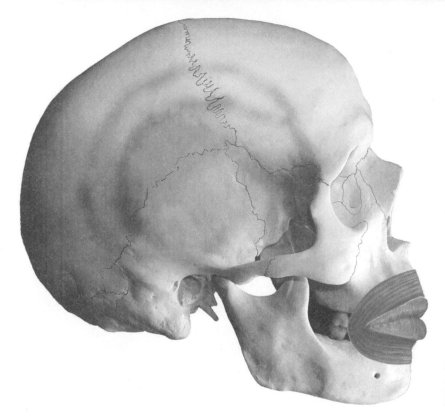

Muscle of facial expression

Skull—lateral view

■ **Origin**

Lateral band—alveolar border of maxilla

Medial band—septum of nose

Inferior portion—lateral to midline of mandible

■ **Insertion**

Becomes continuous with other muscles at angle of mouth

■ **Action**

Closure and protrusion of lips

■ **Nerve**

Buccal and mandibular branches of facial nerve

LEVATOR LABII SUPERIORIS

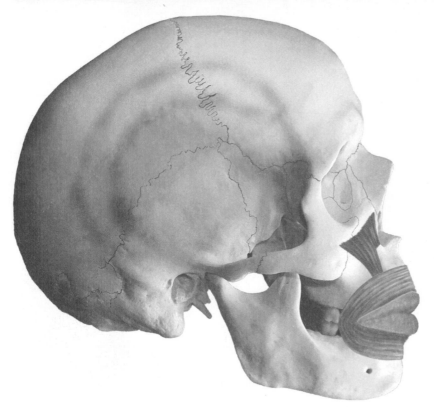

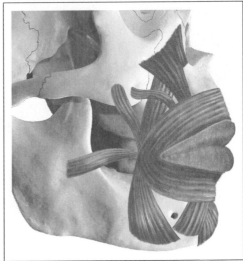

Muscle of facial expression

Skull—lateral view

Angular head *(aleque nasi)*

■ **Origin**
Frontal process of maxilla and zygomatic bone

■ **Insertion**
Greater alar cartilage and skin of nose, upper lip

■ **Action**
Elevates upper lip, dilates nares, forms nasolabial furrow

■ **Nerve**
Buccal branches of facial nerve

Note: The angular head is frequently referred to as a separate muscle, levator labii superioris aleque nasi.

Infraorbital head

■ **Origin**
Infraorbital margin of maxilla

■ **Insertion**
Skin of lateral half of upper lip

■ **Action**
Elevates upper lip, forms nasolabial furrow

■ **Nerve**
Buccal branches of facial nerve

LEVATOR ANGULI ORIS

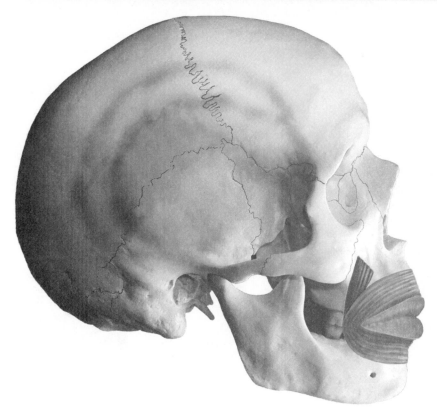

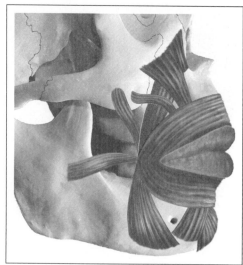

Muscle of facial expression

Skull—lateral view

■ Origin
Canine fossa of maxilla

■ Insertion
Angle of mouth

■ Action
Elevates corner (angle) of mouth

■ Nerve
Buccal branches of facial nerve

ZYGOMATICUS MAJOR

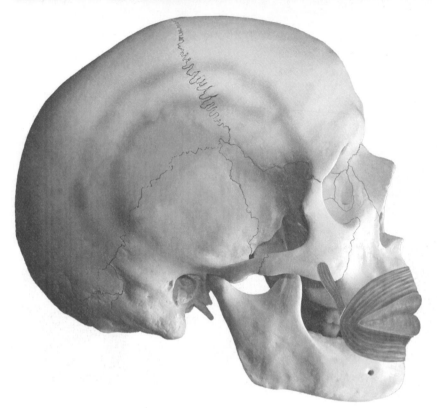

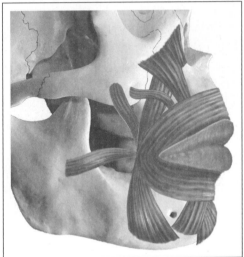

Muscle of facial expression

Skull—lateral view

■ **Origin**
Zygomatic bone

■ **Insertion**
Angle of mouth

■ **Action**
Draws angle of mouth upward and backward (laughing)

■ **Nerve**
Buccal branches of facial nerve

ZYGOMATICUS MINOR

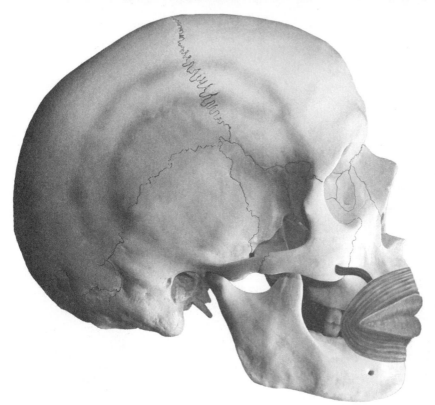

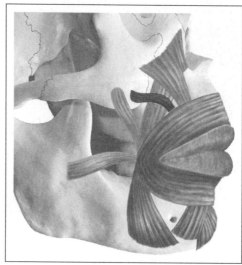

Muscle of facial expression

Skull—lateral view

■ Origin
Zygomatic bone

■ Insertion
Upper lip lateral to levator labii superioris

■ Action
Forms nasolabial furrow

■ Nerve
Buccal branches of facial nerve

RISORIUS

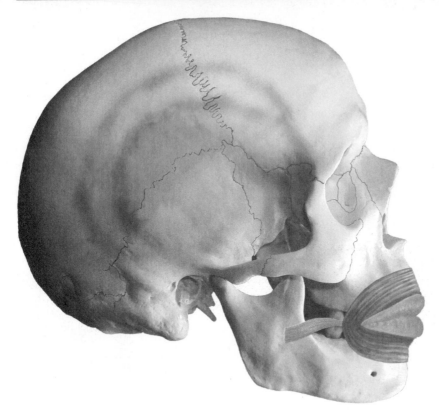

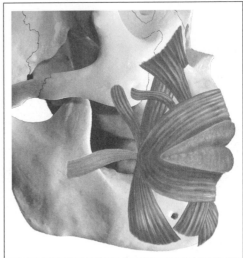

Muscle of facial expression

Skull—lateral view

■ **Origin**
Fascia over masseter

■ **Insertion**
Skin at angle of mouth

■ **Action**
Retracts angle of mouth, as in grinning

■ **Nerve**
Buccal branches of facial nerve

DEPRESSOR LABII INFERIORIS

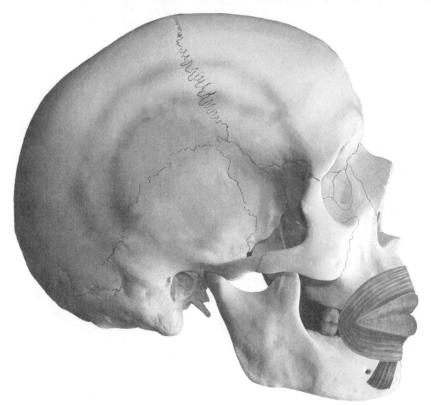

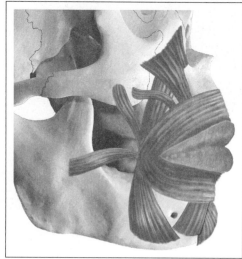

Muscle of facial expression

Skull—lateral view

■ Origin
Mandible, between symphysis and mental foramen

■ Insertion
Skin of lower lip

■ Action
Draws lower lip downward and laterally

■ Nerve
Mandibular branch of facial nerve

DEPRESSOR ANGULI ORIS

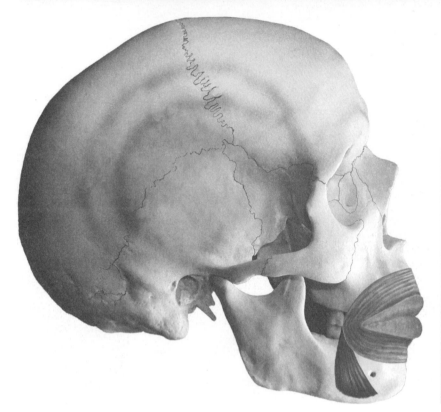

Muscle of facial expression

Skull—lateral view

■ Origin
Oblique line of the mandible

■ Insertion
Angle of the mouth

■ Action
Depresses angle of mouth, as in frowning

■ Nerve
Mandibular branch of facial nerve

MENTALIS

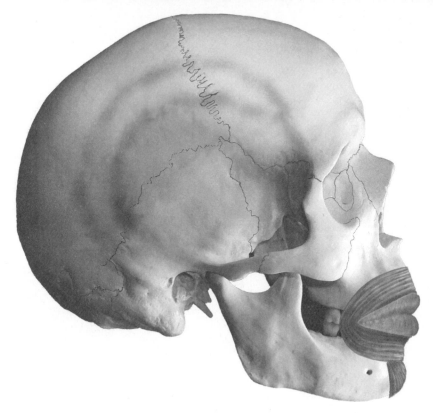

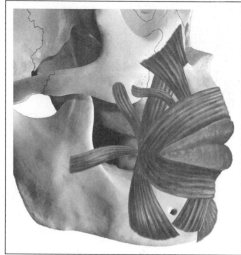

Muscle of facial expression

Skull—lateral view

■ **Origin**
Incisive fossa of mandible

■ **Insertion**
Skin of chin

■ **Action**
Raises and protrudes lower lip, wrinkles skin of chin

■ **Nerve**
Mandibular branch of facial nerve

BUCCINATOR

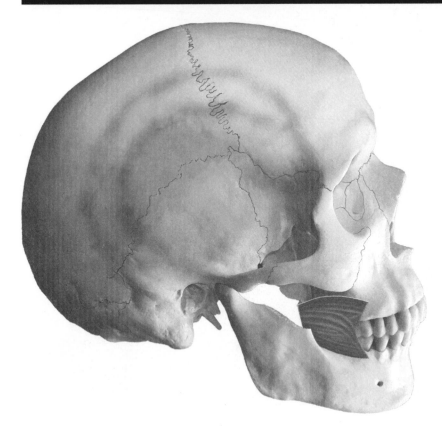

Muscle of facial expression

Skull—lateral view

■ Origin
Outer surface of alveolar processes of maxilla and mandible over molars and along pterygomandibular raphe

■ Insertion
Deep part of muscles of lips

■ Action
Compresses cheek

■ Nerve
Buccal branches of facial nerve

TEMPORALIS

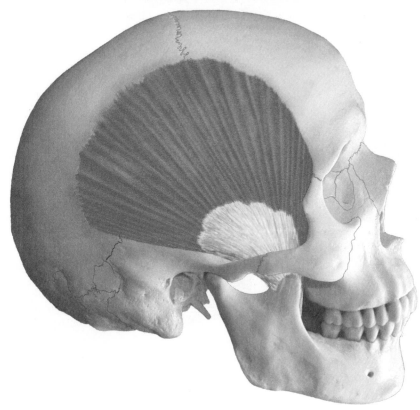

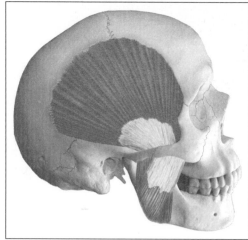

Muscles of mastication

Skull—lateral view

■ Origin
Temporal fossa including frontal, parietal, and temporal bones

■ Insertion
Coronoid process and anterior border of ramus of mandible

■ Action
Elevates, retracts mandible (rotation at temperomandibular joint) closing mouth, biting

■ Nerve
Mandibular division of trigeminal nerve

MASSETER

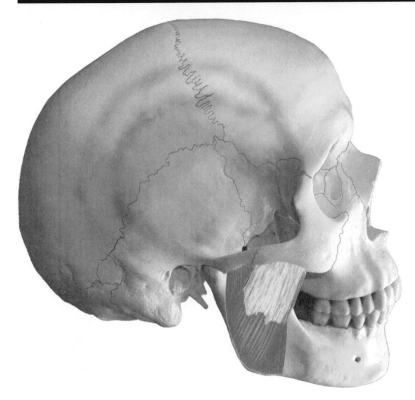

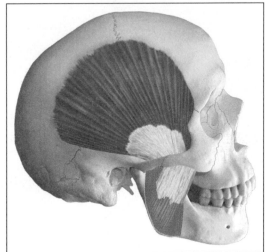

Muscles of mastication

Skull—lateral view

■ Origin
Zygomatic process of maxilla, medial and inferior surfaces of zygomatic arch

■ Insertion
Angle and ramus of mandible, lateral surface of coronoid process of mandible

■ Action
Elevates mandible (rotation at temperomandibular joint) closing mouth, biting

■ Nerve
Mandibular division of trigeminal nerve

Note: Superficial fibers slightly protract jaw (see lateral pterygoid).

MEDIAL PTERYGOID *(Pterygoideus Medialis)*

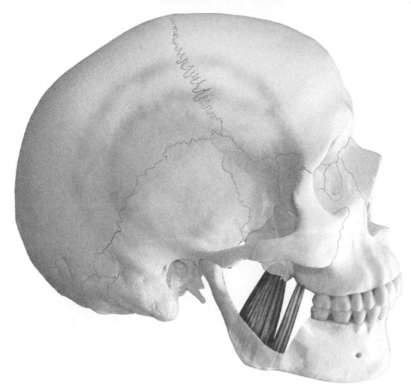

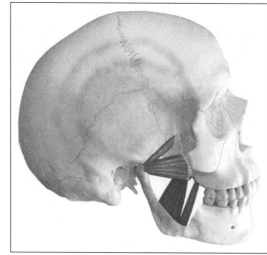

Muscles of mastication

Skull—lateral view
(Part of mandible cut away)

■ Origin
Medial surface of lateral pterygoid plate of sphenoid bone, palatine bone, and tuberosity of maxilla

■ Insertion
Medial surface of ramus and angle of mandible

■ Action
Elevates mandible, assists in protruding mandible

■ Nerve
Mandibular division of trigeminal nerve

LATERAL PTERYGOID *(Pterygoideus Lateralis)*

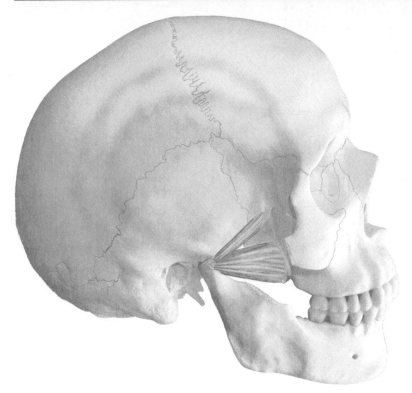

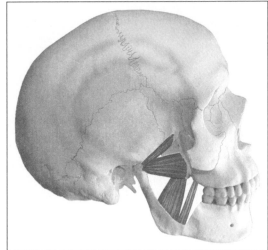

Muscles of mastication

Skull—lateral view

■ Origin
Superior head—lateral surface of greater wing of sphenoid

Inferior head—lateral surface of lateral pterygoid plate

■ Insertion
Condyle of mandible, temporomandibular joint

■ Action
Opens jaw, protrudes mandible, moves mandible laterally for grinding teeth

■ Nerve
Mandibular division of trigeminal nerve

Note: This sideward movement, aided by superficial fibers of masseter, causes chewing movements.

CHAPTER FOUR

Muscles of the Neck

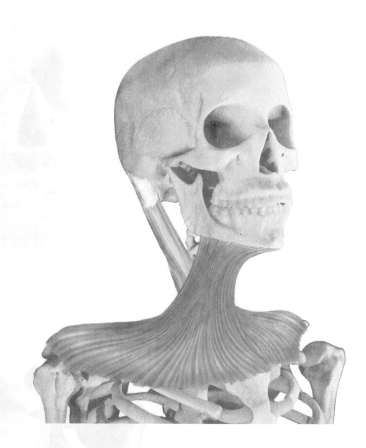

STERNOCLEIDOMASTOID

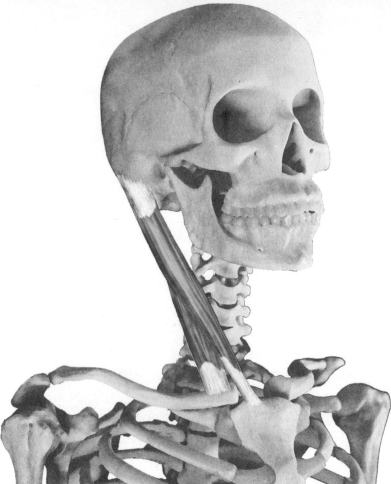

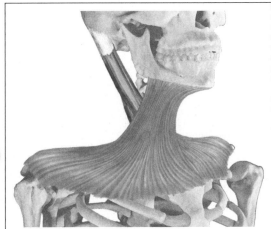

Superficial muscles of the neck

Three-quarter frontal view

■ Origin
Sternal head—manubrium of sternum

Clavicular head—medial part of clavicle

■ Insertion
Mastoid process of temporal bone, lateral half
of superior nuchal line of occipital bone

■ Action
Unilateral—bends neck laterally, rotates head
to opposite side

Bilateral—flexes neck, draws head ventrally and elevates
chin, draws sternum superiorly in deep inspiration

■ Nerve
Spinal part of accessory nerve (C2, C3)

PLATYSMA

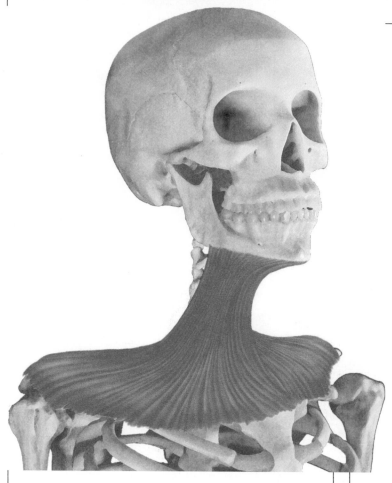

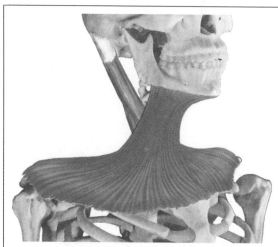

Superficial muscles of the neck

Three-quarter frontal view

■ **Origin**
Subcutaneous fascia of upper one-fourth of chest just below the clavicle

■ **Insertion**
Subcutaneous fascia and muscles of chin and jaw, mandible

■ **Action**
Depresses and draws lower lip laterally, draws up skin of chest, depresses mandible

■ **Nerve**
Cervical branch of facial nerve

DIGASTRICUS

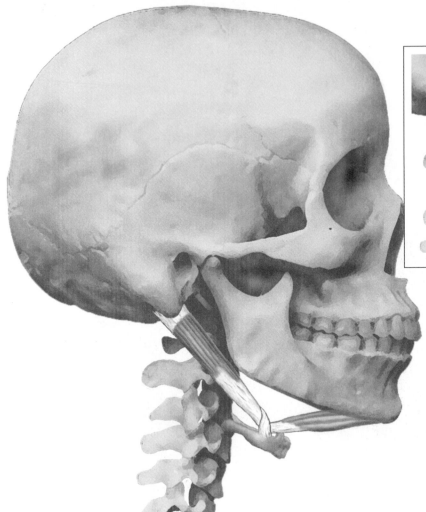

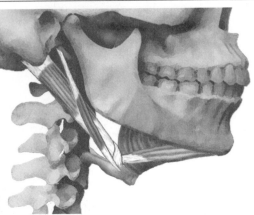

Suprahyoid muscles

Lateral view

■ Origin
Posterior belly—mastoid notch of temporal bone

Anterior belly—inner side of inferior border of mandible near symphysis

■ Insertion
Intermediate tendon attached to hyoid bone

■ Action
Raises hyoid bone, depresses mandible, moves hyoid forward or backward

■ Nerve
Posterior belly—facial nerve

Anterior belly—mandibular division of trigeminal

Note: The two bellies are joined at an intermediate tendon by a fibrous loop at the side of the body and the greater horn of the hyoid bone.

STYLOHYOID

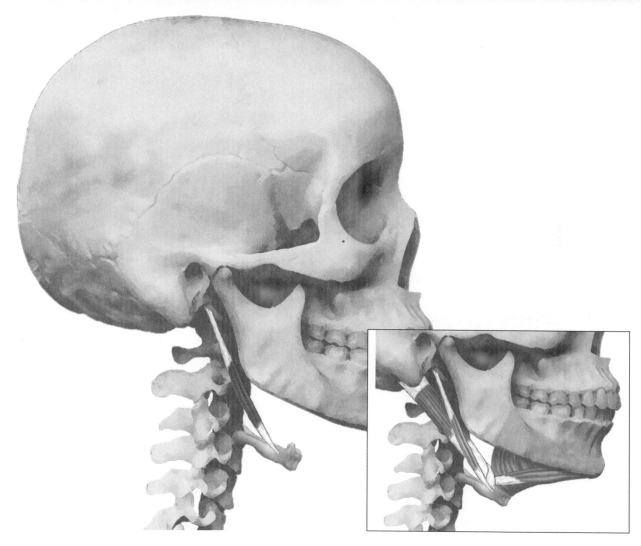

Suprahyoid muscles

Lateral view

■ Origin
Styloid process of temporal bone

■ Insertion
Hyoid bone

■ Action
Elevates and draws hyoid bone posteriorly, elevates tongue

■ Nerve
Facial nerve

MYLOHYOID

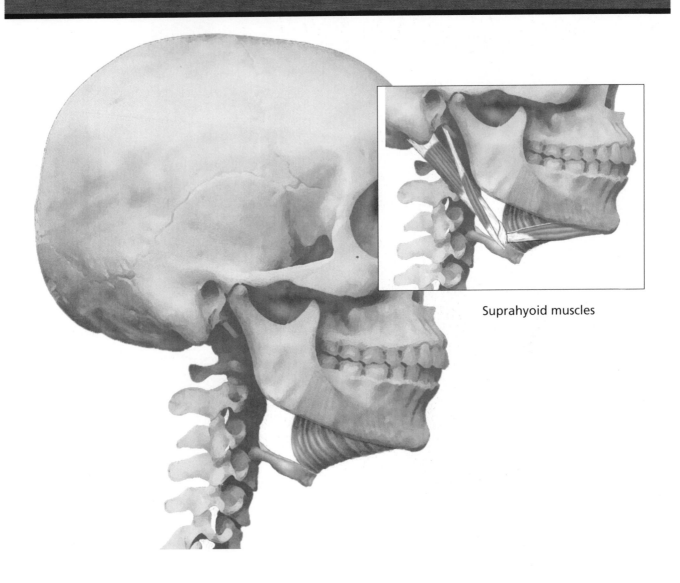

Suprahyoid muscles

Lateral view

■ **Origin**
Inside surface of mandible from symphysis to molars (mylohyoid line)

■ **Insertion**
Hyoid bone

■ **Action**
Elevates hyoid bone, raises floor of mouth and tongue

■ **Nerve**
Mandibular division of trigeminal nerve

GENIOHYOID

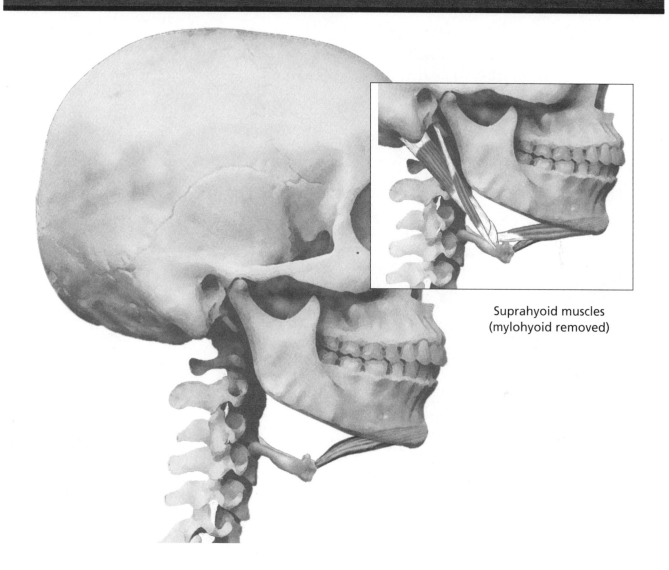

Suprahyoid muscles
(mylohyoid removed)

Lateral view

■ Origin
Inferior mental spine on interior medial surface of mandible

■ Insertion
Body of hyoid bone

■ Action
Protrudes hyoid bone and tongue

■ Nerve
Branch of C1 through hypoglossal nerve

STERNOHYOID

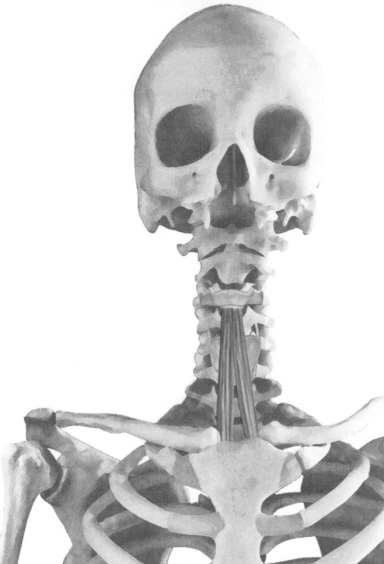

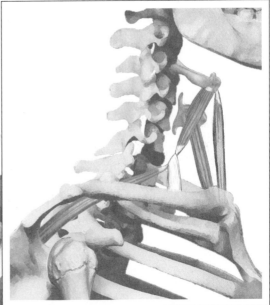

Infrahyoid muscles

Frontal view
(Mandible and part of maxilla removed)

■ **Origin**
Medial end of clavicle, manubrium of sternum

■ **Insertion**
Body of hyoid bone

■ **Action**
Depresses hyoid bone

■ **Nerve**
Ansa cervicalis (C1–C3)

Lateral view

STERNOTHYROID

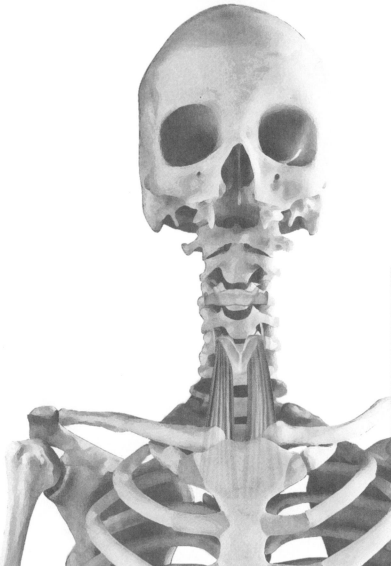

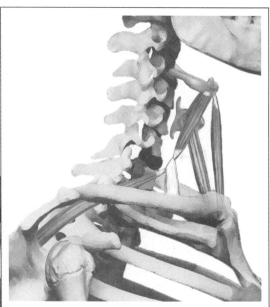

Infrahyoid muscles

Frontal view
(Mandible and part of maxilla removed)

■ Origin
Dorsal surface of manubrium of sternum

■ Insertion
Lamina of thyroid cartilage

■ Action
Depresses thyroid cartilage

■ Nerve
Ansa cervicalis (C1–C3)

Lateral view

THYROHYOID

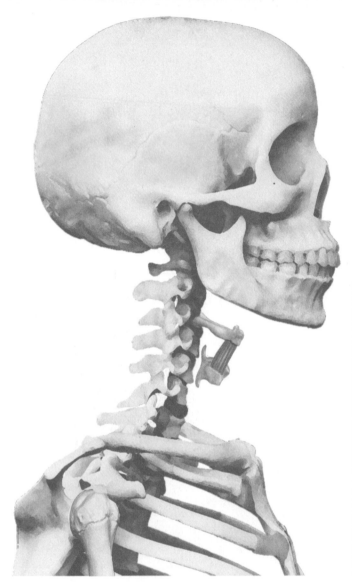

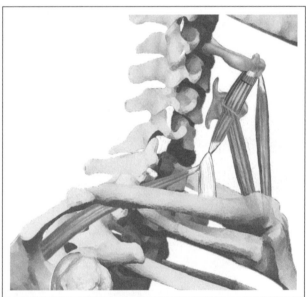

Infrahyoid muscles (superior omohyoid cut)

Lateral view

■ Origin
Lamina of thyroid cartilage

■ Insertion
Inferior border of body and greater horn of hyoid bone

■ Action
Depresses hyoid or raises thyroid

■ Nerve
C1 through hypoglossal nerve

OMOHYOID

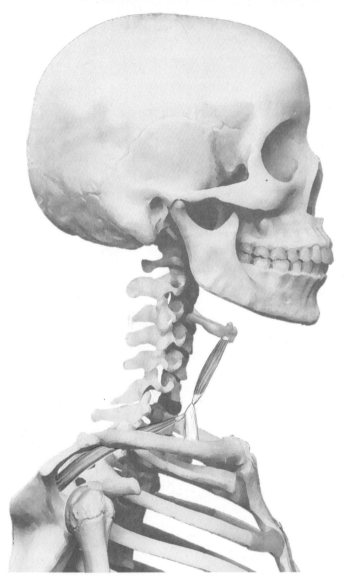

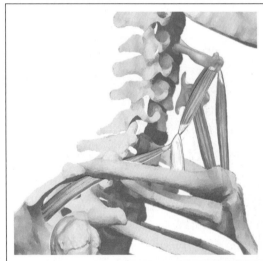

Infrahyoid muscles

Lateral view

■ Origin
Superior border of scapula

■ Insertion
Inferior belly—bound to clavicle by central tendon

Superior belly—continues to body of hyoid bone

■ Action
Depresses hyoid bone

■ Nerve
Ansa cervicalis (C2, C3)

LONGUS COLLI

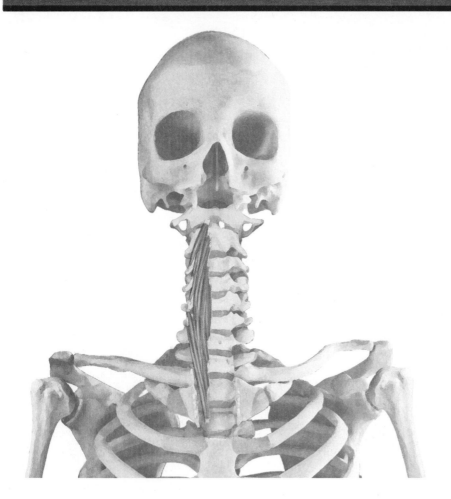

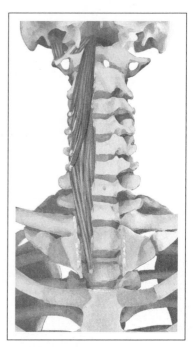

Anterior muscles
(manubrium removed)

Frontal view
(Mandible, part of maxilla, and manubrium of sternum removed)

Superior oblique part

■ **Origin**
Transverse processes of third, fourth, and fifth cervical vertebrae

■ **Insertion**
Anterior arch of atlas

Inferior oblique part

■ **Origin**
Anterior surface of bodies of first two or three thoracic vertebrae

■ **Insertion**
Transverse processes of fifth and sixth cervical vertebrae

Vertical part

■ **Origin**
Anterior surfaces of bodies of upper three thoracic and lower three cervical vertebrae

■ **Insertion**
Anterior surfaces of the second, third, and fourth cervical vertebrae

■ **Action**
All three parts flex cervical vertebrae

■ **Nerve**
C2–C7

Note: Cervical hyperextension injuries (whiplash) may strain these muscles and/or sprain the anterior ligaments of vertebrae.

LONGUS CAPITIS

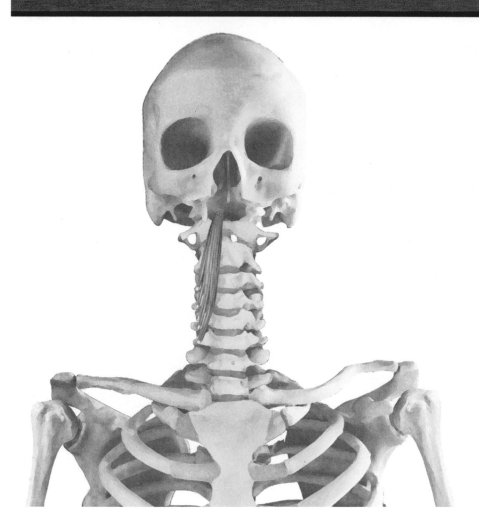

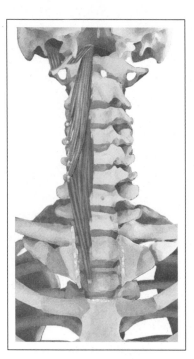

Anterior muscles
(manubrium removed)

Frontal view
(Mandible and part of maxilla removed)

■ **Origin**
Anterior tubercles of transverse processes of third through sixth cervical vertebrae

■ **Insertion**
Occipital bone anterior to foramen magnum

■ **Action**
Acting together (bilaterally)—flex head

Acting on one side only—rotate head

■ **Nerve**
C1–C3

RECTUS CAPITIS ANTERIOR

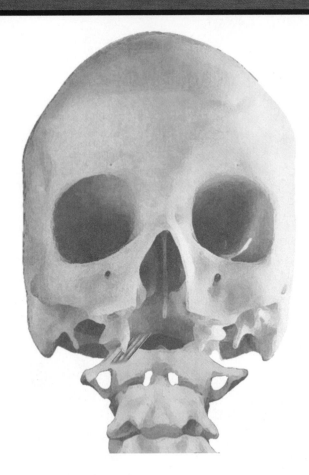

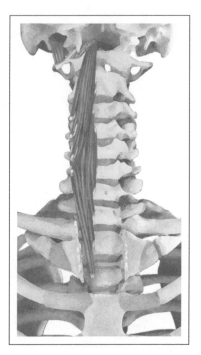

Anterior muscles
(manubrium removed)

Frontal view
(Mandible and part of maxilla removed)

■ **Origin**
Anterior base of transverse process of atlas

■ **Insertion**
Occipital bone anterior to foramen magnum

■ **Action**
Flexes head

■ **Nerve**
C1, C2

RECTUS CAPITIS LATERALIS

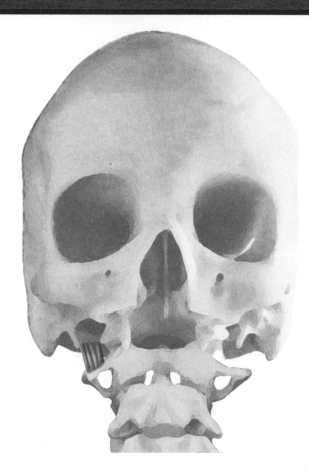

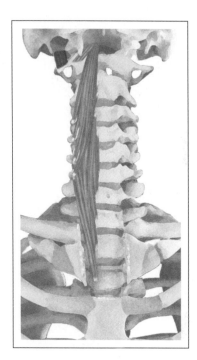

Anterior muscles
(manubrium removed)

Frontal view
(Mandible and part of maxilla removed)

■ **Origin**
Transverse process of atlas

■ **Insertion**
Jugular process of occipital bone

■ **Action**
Bends head laterally

■ **Nerve**
C1, C2

SCALENUS ANTERIOR

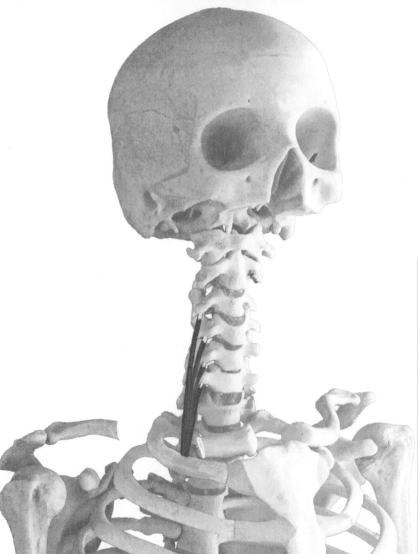

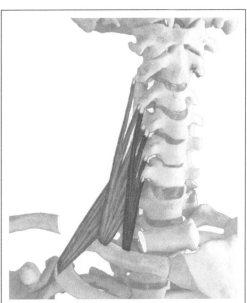

Lateral muscles

Three-quarter frontal view
(Mandible, part of maxilla, and part of clavicle removed)

■ Origin
Transverse processes of third through sixth cervical vertebrae

■ Insertion
Inner border of first rib (scalene tubercle)

■ Action
Raises first rib (respiratory inspiration); acting together, they flex neck; acting on one side, they laterally flex, rotate neck

■ Nerve
Ventral rami of cervical nerves (C4–C6)

SCALENUS MEDIUS

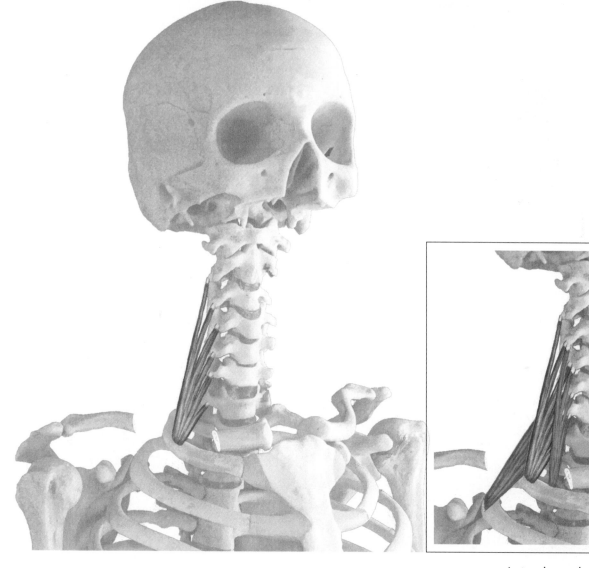

Lateral muscles

Three-quarter frontal view
(Mandible, part of maxilla, and part of clavicle removed)

■ Origin
Transverse processes of lower six cervical vertebrae (C2–C7)

■ Insertion
Upper surface of first rib

■ Action
Raises first rib (respiratory inspiration); acting together, they flex neck; acting on one side, they laterally flex, rotate neck

■ Nerve
Ventral rami of cervical nerves (C3–C8)

SCALENUS POSTERIOR

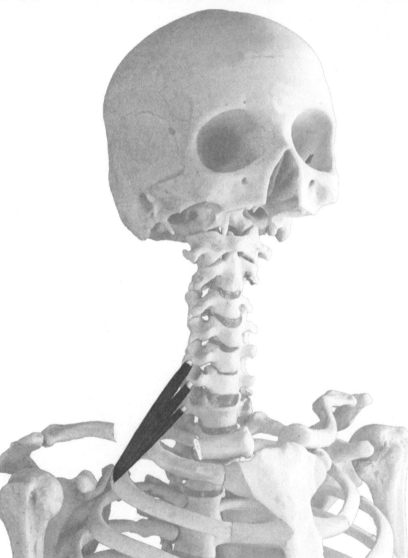

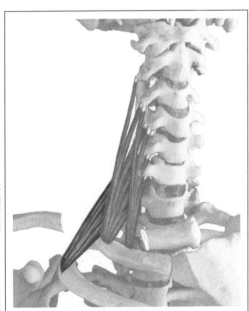

Lateral muscles

Three-quarter frontal view
(Mandible and part of maxilla removed)

■ Origin
Transverse processes of lower two or three cervical
vertebrae (C5–C7)

■ Insertion
Outer surface of second rib

■ Action
Raises second rib (respiratory inspiration); acting
together, they flex neck; acting on one side, they
laterally flex, rotate neck

■ Nerve
Ventral rami of lower three cervical nerves

RECTUS CAPITIS POSTERIOR MAJOR

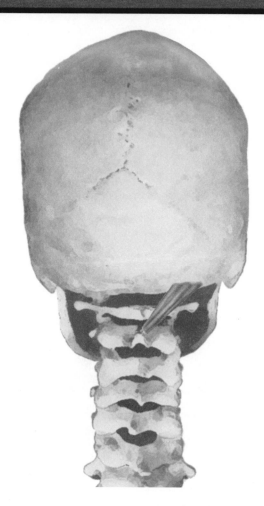

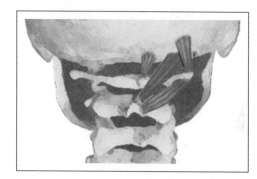

Deep posterior (suboccipital) muscles

Posterior skull and cervical vertebrae

■ Origin
Spinous process of axis

■ Insertion
Lateral portion of inferior nuchal line of occipital bone

■ Action
Extends and rotates head

■ Nerve
Suboccipital nerve

RECTUS CAPITIS POSTERIOR MINOR

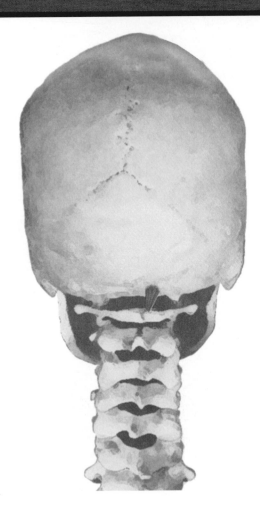

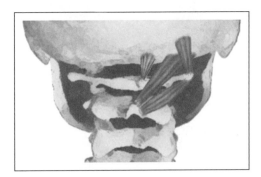

Deep posterior (suboccipital) muscles

Posterior skull and cervical vertebrae

■ **Origin**
Posterior arch of atlas

■ **Insertion**
Medial portion of inferior nuchal line of occipital bone

■ **Action**
Extends head

■ **Nerve**
Suboccipital nerve

OBLIQUUS CAPITIS INFERIOR

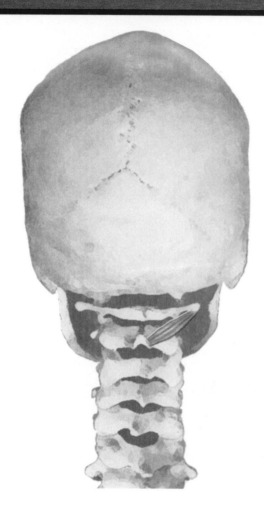

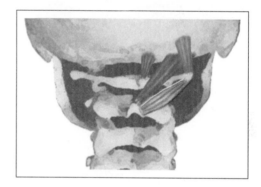

Deep posterior (suboccipital) muscles

Posterior skull and cervical vertebrae

■ **Origin**
Spinous process of axis

■ **Insertion**
Transverse process of atlas

■ **Action**
Rotates atlas

■ **Nerve**
Suboccipital nerve

OBLIQUUS CAPITIS SUPERIOR

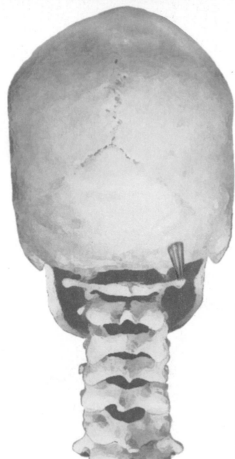

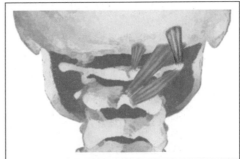

Deep posterior (suboccipital) muscles

Posterior skull and cervical vertebrae

■ **Origin**
Transverse process of atlas

■ **Insertion**
Occipital bone between inferior and superior
nuchal lines

■ **Action**
Extends head and flexes head laterally to the same side
(ipsilaterally)

■ **Nerve**
Suboccipital nerve

CHAPTER FIVE

Muscles of the Trunk

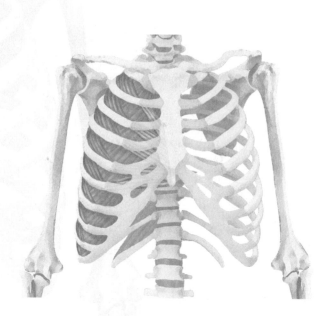

SPLENIUS CAPITIS

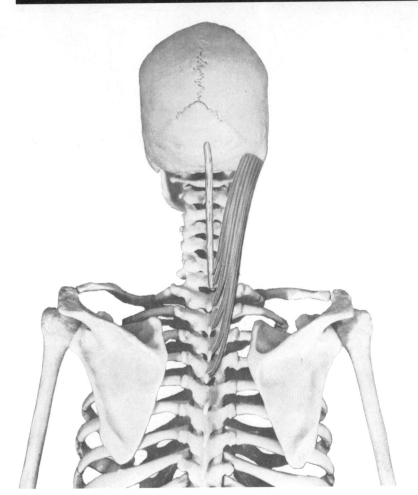

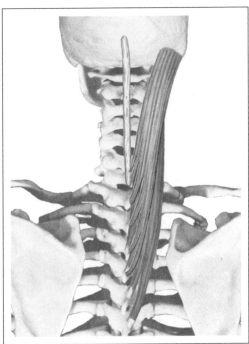

Superficial muscles of neck
and upper thorax

Posterior skull, neck, and back

■ Origin
Lower part of ligamentum nuchae, spinous processes of seventh cervical vertebra (C7) and upper three or four thoracic vertebrae (T1–T4)

■ Insertion
Mastoid process of temporal bone and lateral part of superior nuchal line

■ Action
Acting together, they extend, hyperextend head, neck; acting on one side, they laterally flex, rotate head, neck

■ Nerve
Lateral branches of dorsal primary divisions of middle and lower cervical nerves (C2, C3)

■ Relationship
This muscle is deep to trapezius and superficial to semispinalis capitis and longissimus capitus

SPLENIUS CERVICIS

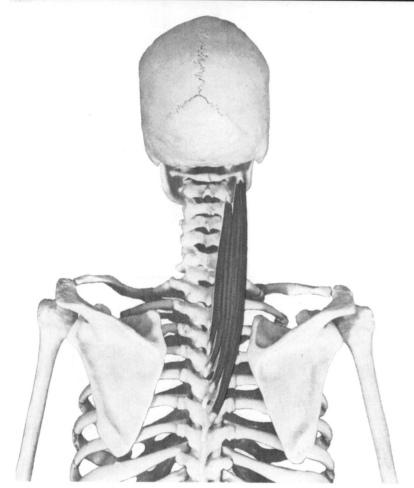

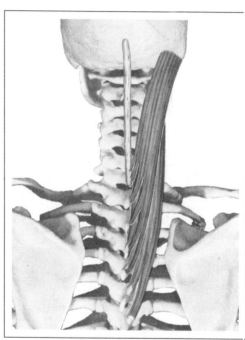

Superficial muscles of neck
and upper thorax

Posterior skull, neck, and back

■ Origin
Spinous processes of third through sixth thoracic
vertebrae (T3–T6)

■ Insertion
Transverse processes of upper two or three cervical
vertebrae (C1–C3)

■ Action
Acting together, they extend, hyperextend head, neck;
acting on one side, they laterally flex, rotate head, neck

■ Nerve
Lateral branches of dorsal primary divisions of middle
and lower cervical nerves

■ Relationship
This muscle is deep to serratus posterior superior,
rhomboids, and trapezius

ERECTOR SPINAE

Iliocostalis cervicis

■ **Origin**
Angles of third through sixth ribs

■ **Insertion**
Transverse processes of fourth, fifth, and sixth cervical vertebrae

■ **Action**
Extension, lateral flexion of vertebral column

■ **Nerve**
Dorsal primary divisions of spinal nerves

Iliocostalis thoracis

■ **Origin**
Angles of lower six ribs medial to iliocostalis lumborum

■ **Insertion**
Angles of upper six ribs and transverse process of seventh cervical vertebra

■ **Action**
Extension, lateral flexion of vertebral column, rotates ribs for forceful inspiration

■ **Nerve**
Dorsal primary divisions of spinal nerves

Iliocostalis lumborum

■ **Origin**
Medial and lateral sacral crests and medial part of iliac crests

■ **Insertion**
Angles of lower six ribs

■ **Action**
Extension, lateral flexion of vertebral column, rotates ribs for forceful inspiration

■ **Nerve**
Dorsal primary divisions of spinal nerves

Note: The erector spinae (sacrospinalis) is a complex of three sets of muscles: iliocostalis, longissimus, and spinalis. The origin of this group is the medial and lateral sacral crests, the medial part of iliac crests, and the spinous processes and supraspinal ligament of lumbar and eleventh and twelfth thoracic vertebrae.

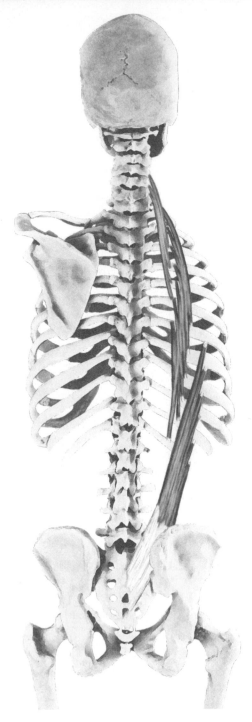

Trunk—dorsal view

ERECTOR SPINAE

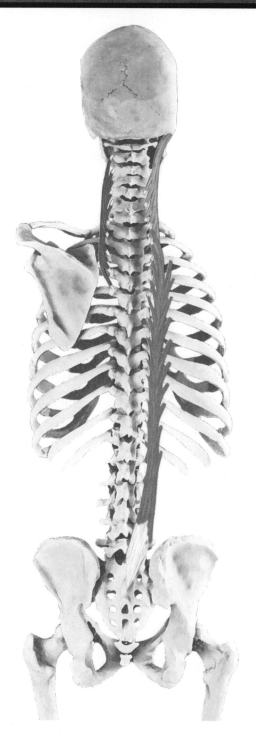

Trunk—dorsal view

Longissimus capitis

■ Origin
Transverse processes of upper five thoracic vertebrae (T1–T5), articular processes of lower three cervical vertebrae (C5–C7)

■ Insertion
Posterior part of mastoid process of temporal bone

■ Action
Extends and rotates head

■ Nerve
Dorsal primary divisions of middle and lower cervical nerves

Longissimus cervicis

■ Origin
Transverse processes of upper four or five thoracic vertebrae (T1–T5)

■ Insertion
Transverse processes of second through sixth cervical vertebrae (C2–C6)

■ Action
Extension, lateral flexion of vertebral column

■ Nerve
Dorsal primary divisions of spinal nerves

Longissimus thoracis

■ Origin
Medial and lateral sacral crests, spinous processes and supraspinal ligament of lumbar and eleventh and twelfth thoracic vertebrae, and medial part of iliac crests

■ Insertion
Transverse processes of all thoracic vertebrae, between tubercles and angles of lower nine or ten ribs

■ Action
Extension, lateral flexion of vertebral column, rotates ribs for forceful inspiration

■ Nerve
Dorsal primary divisions of spinal nerves

ERECTOR SPINAE

Spinalis capitis
(Medial part of semispinalis capitis)

Spinalis cervicis

■ Origin
Ligamentum nuchae, spinous process of seventh cervical vertebra

■ Insertion
Spinous process of axis

■ Action
Extends vertebral column

■ Nerve
Dorsal primary divisions of spinal nerves

Spinalis thoracis

■ Origin
Spinous processes of lower two thoracic (T11, T12) and upper two lumbar (L1, L2) vertebrae

■ Insertion
Spinous processes of upper thoracic vertebrae (T1–T8)

■ Action
Extends vertebral column

■ Nerve
Dorsal primary divisions of spinal nerves

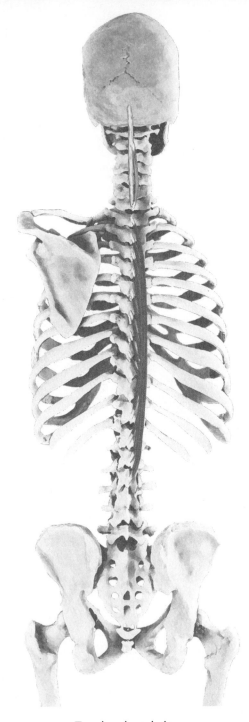

Trunk—dorsal view

ERECTOR SPINAE

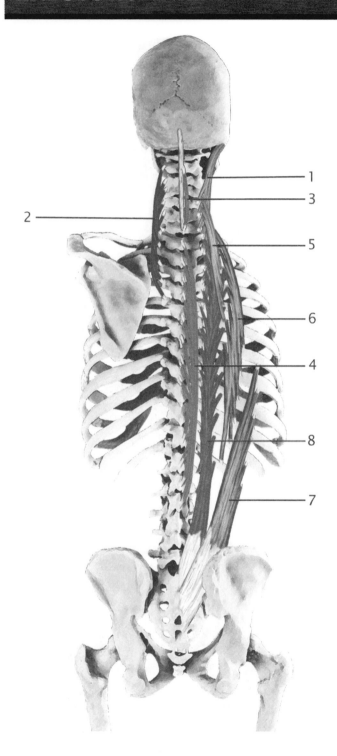

1. Longissimus capitis
2. Longissimus cervicis
3. Spinalis cervicis
4. Spinalis thoracis
5. Iliocostalis cervicis
6. Iliocostalis thoracis
7. Iliocostalis lumborum
8. Longissimus thoracis

ROTATORES*

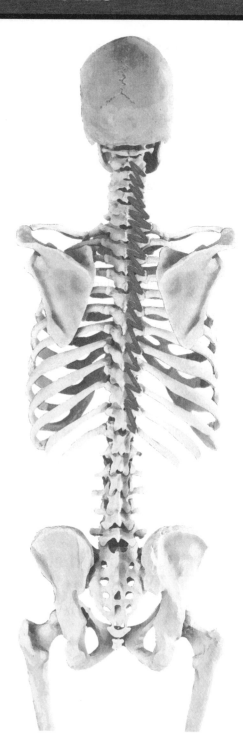

Enlargement of T1–T2 vertebrae

Trunk—dorsal view

■ Origin
Transverse process of each vertebra

■ Insertion
Base of spinous process of next vertebra above

■ Action
Extend and rotate vertebral column

■ Nerve
Dorsal primary division of spinal nerves

*Part of transversospinalis.

INTERSPINALES *(Paired on either side of interspinal ligament)*

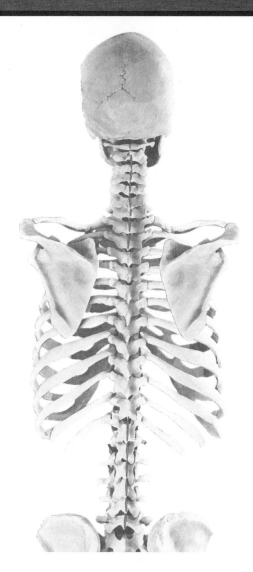

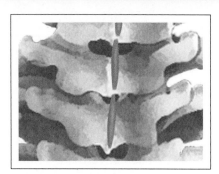

Enlargement of T1–T2 vertebrae

Trunk—dorsal view

■ Origin

Cervical region—spinous processes of third to seventh cervical vertebrae (C3–C7)

Thoracic region—spinous processes of first to third (T1–T3) and eleventh and twelfth thoracic vertebrae (T11, T12)

Lumbar region—spinous processes of second to fifth lumbar vertebrae (L2–L5)

■ Insertion

Spinous process of next vertebra superior to origin

■ Action

Extend vertebral column

■ Nerve

Dorsal primary division of spinal nerves

INTERTRANSVERSARII

Cervical region

Intertransversarii anteriores

- **Origin** Anterior tubercle of transverse processes of vertebrae from first thoracic to axis
- **Insertion** Anterior tubercle of next superior vertebra
- **Action** Lateral flexion of vertebral column
- **Nerve** Ventral primary division of spinal nerves

Intertransversarii posteriores

- **Origin** Posterior tubercle of transverse processes of vertebrae from first thoracic to axis
- **Insertion** Posterior tubercle of next superior vertebra

Thoracic region

- **Origin** Transverse processes of first lumbar to eleventh thoracic vertebrae
- **Insertion** Transverse processes of next superior vertebra

Lumbar region

Intertransversarii laterales

- **Origin** Transverse processes of lumbar vertebrae
- **Insertion** Transverse process of next superior vertebra
- **Action** Lateral flexion of vertebral column
- **Nerve** Ventral primary division of spinal nerves

Intertransversarii mediales

- **Origin** Mammillary process* of each lumbar vertebra
- **Insertion** Accessory process of the next superior lumbar vertebra
- **Action** Lateral flexion of vertebral column
- **Nerve** Dorsal primary division of spinal nerves

*Posterior border of superior articular process.

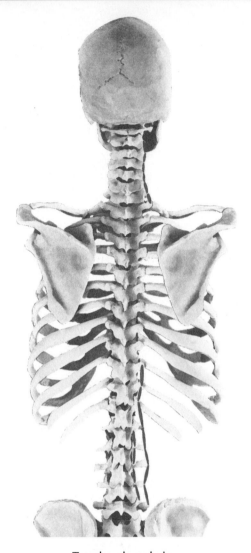

Trunk—dorsal view

Enlargement of L4–L5 vertebrae

EXTERNAL INTERCOSTALS *(Intercostales Externi)*

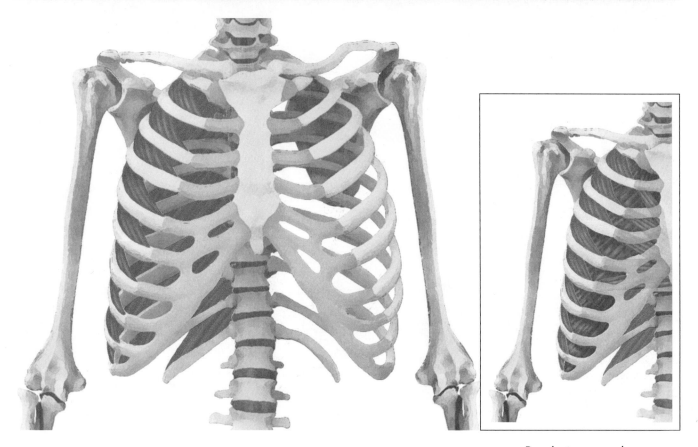

Respiratory muscles

Trunk—anterior view

■ **Origin**
Lower margin of upper eleven ribs

■ **Insertion**
Superior border of rib below (each muscle fiber runs obliquely and inserts toward the costal cartilage)

■ **Action**
Draw ventral part of ribs upward, increasing the volume of the thoracic cavity for inspiration

■ **Nerve**
Intercostal nerves

INTERNAL INTERCOSTALS *(Intercostales Interni)*

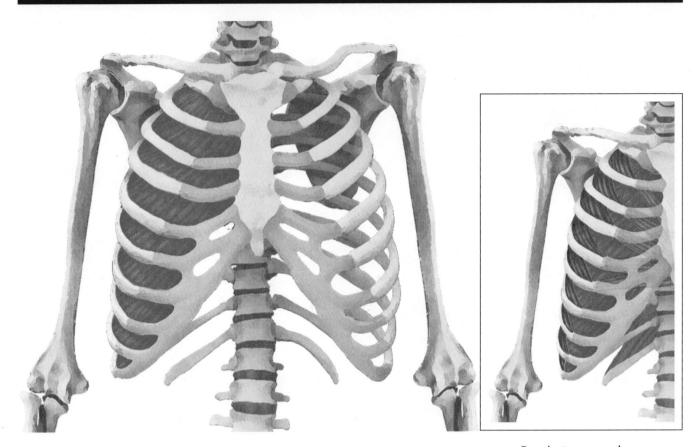

Respiratory muscles

Trunk—anterior view

■ Origin
From the cartilages to the angles of the upper eleven ribs

■ Insertion
Superior border of the rib below (each muscle fiber runs obliquely and inserts away from the costal cartilage)

■ Action
Draw ventral part of ribs downward, decreasing the volume of the thoracic cavity for expiration

■ Nerve
Intercostal nerves

SUBCOSTALS *(Subcostales)*

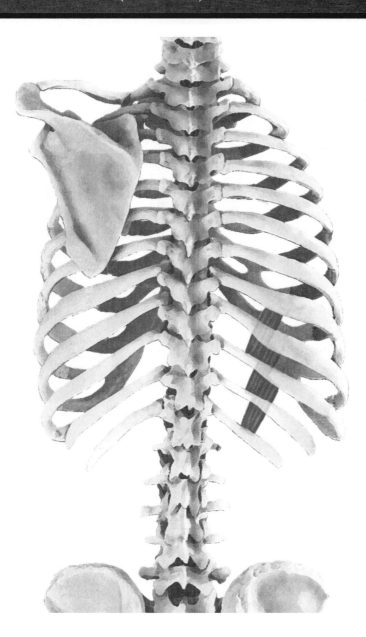

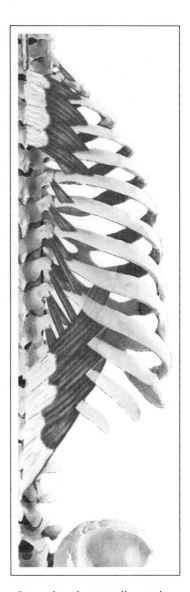

Posterior chest wall muscles

Trunk—dorsal view

■ Origin
Inner surface of each rib near its angle

■ Insertion
Medially on the inner surface of second or third rib below

■ Action
Draw ventral part of ribs downward, decreasing the volume of the thoracic cavity for forceful expiration

■ Nerve
Intercostal nerves

Note: These muscles are deep to the internal intercostals. They continue distally between single ribs, where they are known as innermost intercostal muscles.

SERRATUS POSTERIOR SUPERIOR

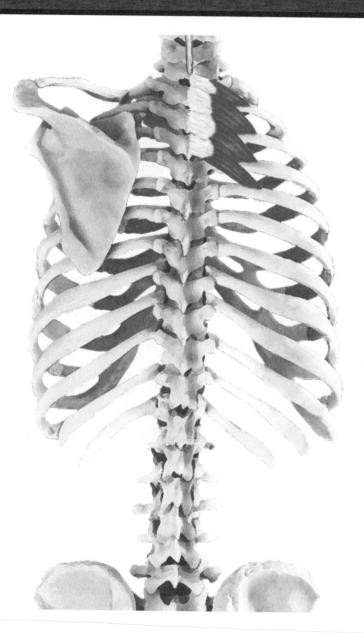

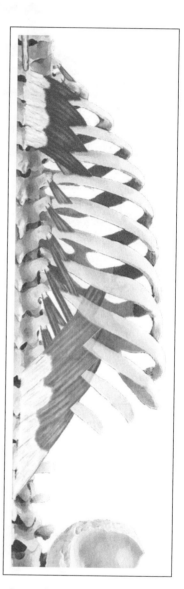

Posterior chest wall muscles

Trunk—dorsal view

■ Origin
Ligamentum nuchae, spinous processes of seventh cervical and first few thoracic vertebrae

■ Insertion
Upper borders of the second through fifth ribs lateral to their angles

■ Action
Raises ribs in inspiration

■ Nerve
T1–T4

SUBCOSTALS *(Subcostales)*

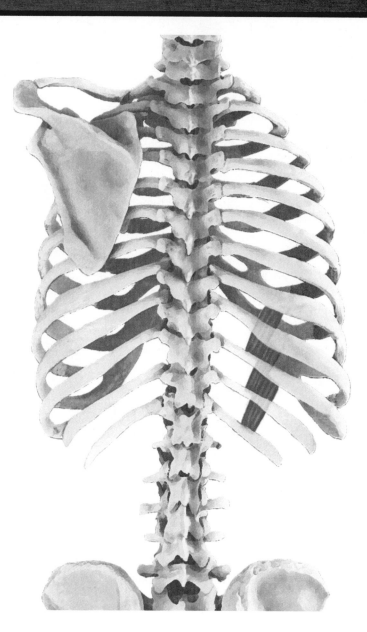

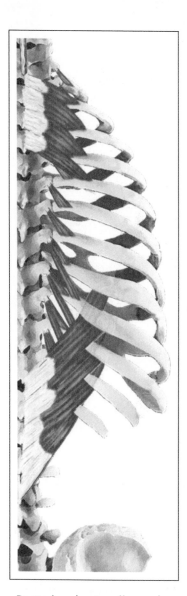

Posterior chest wall muscles

Trunk—dorsal view

■ Origin
Inner surface of each rib near its angle

■ Insertion
Medially on the inner surface of second or third rib below

■ Action
Draw ventral part of ribs downward, decreasing the volume of the thoracic cavity for forceful expiration

■ Nerve
Intercostal nerves

Note: These muscles are deep to the internal intercostals. They continue distally between single ribs, where they are known as innermost intercostal muscles.

TRANSVERSUS THORACIS

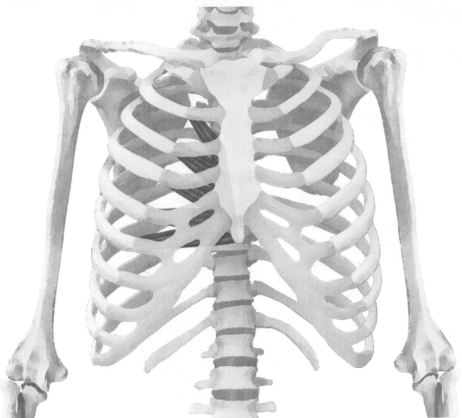

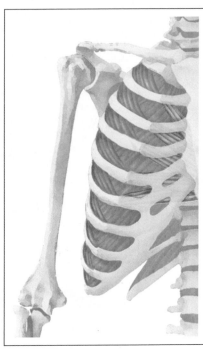

Respiratory muscles

Trunk—anterior view

■ **Origin**
Inner surface of lower portion of sternum and adjacent costal cartilages

■ **Insertion**
Inner surfaces of costal cartilages of the second through sixth ribs

■ **Action**
Draws ventral part of ribs downward, decreasing the volume of the thoracic cavity for forceful expiration

■ **Nerve**
Intercostal nerves

Note: These muscles are deep to the internal intercostal muscles.

LEVATORES COSTARUM

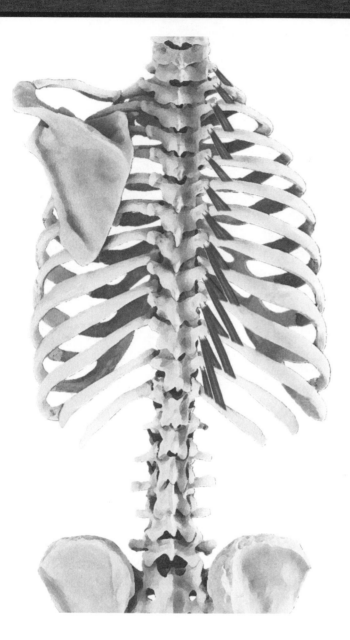

Posterior chest wall muscles

Trunk—dorsal view

■ Origin
Transverse processes of the seventh cervical and the upper eleven thoracic vertebrae

■ Insertion
Laterally to outer surface of next lower rib (lower muscles may cross over one rib)

■ Action
Raises ribs; extends, laterally flexes, and rotates vertebral column

■ Nerve
Intercostal nerves

SERRATUS POSTERIOR SUPERIOR

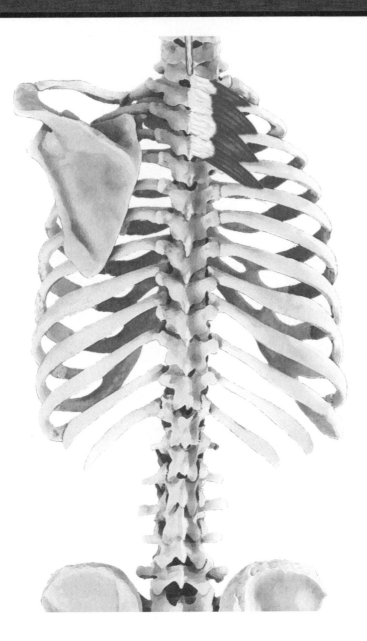

Posterior chest wall muscles

Trunk—dorsal view

■ Origin
Ligamentum nuchae, spinous processes of seventh cervical and first few thoracic vertebrae

■ Insertion
Upper borders of the second through fifth ribs lateral to their angles

■ Action
Raises ribs in inspiration

■ Nerve
T1–T4

SERRATUS POSTERIOR INFERIOR

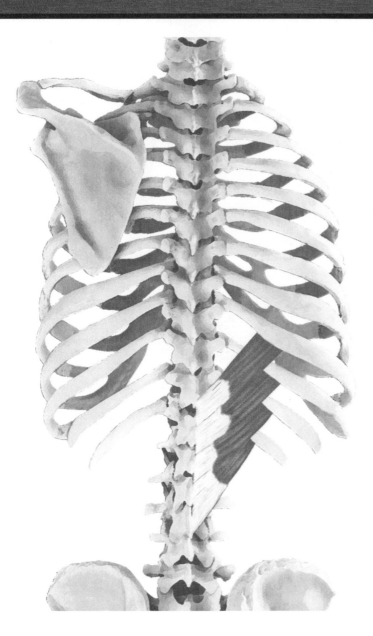

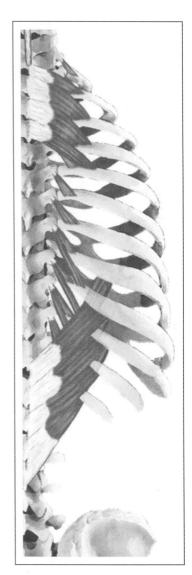

Posterior chest wall muscles

Trunk—dorsal view

■ Origin
Spinous processes of the lower two thoracic and the upper two or three lumbar vertebrae

■ Insertion
Lower borders of bottom four ribs

■ Action
Pulls ribs down, resisting pull of diaphragm

■ Nerve
T9–T12

DIAPHRAGM

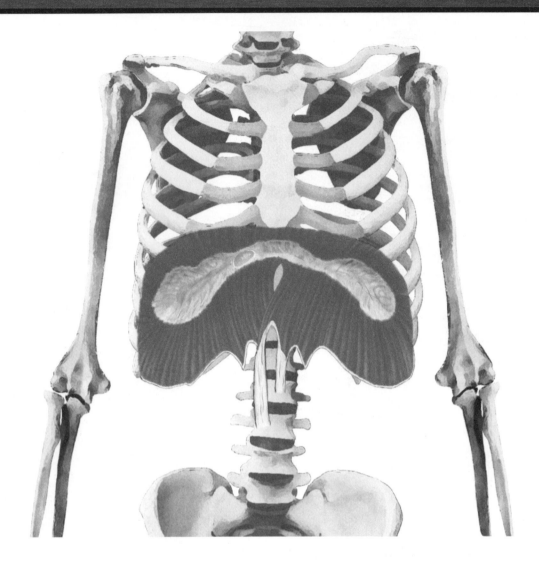

Trunk—anterior view
(Lower costal cartilages removed)

■ Origin
Sternal part—inner part of xiphoid process

Costal part—inner surfaces of lower six ribs and their cartilages

Lumbar part—upper two or three lumbar vertebrae and lateral and medial lumbocostal arches*

■ Insertion
Fibers converge and meet on a central tendon

■ Action
Draws central tendon inferiorly, for inspiration

■ Nerve
Phrenic nerve (C3–C5)

Note: This muscle inserts upon itself. Its action is to change the volume of the thoracic and abdominal cavities.

*These tendinous structures, also known as the medial and lateral arcuate ligaments, allow the diaphragm to bridge the upper parts of the psoas major and quadratus lumborum muscles.

EXTERNAL OBLIQUE *(Obliquus Externus Abdominis)*

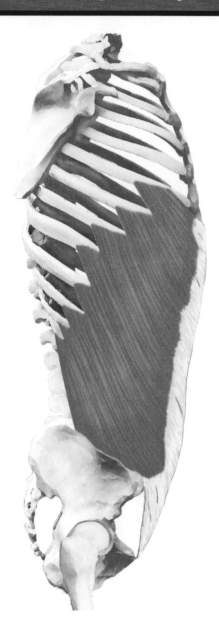

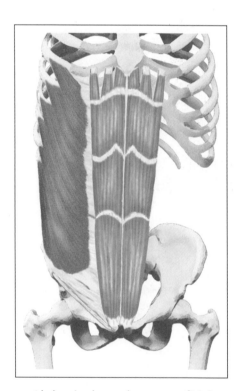

Abdominal muscles—superficial

Trunk—lateral view

■ Origin
Lower eight ribs

■ Insertion
Anterior part of iliac crest, abdominal aponeurosis (rectus sheath) to linea alba

■ Action
Compresses abdominal contents, laterally flexes and rotates vertebral column

■ Nerve
Eighth to twelfth intercostal, iliohypogastric, ilioinguinal nerves

■ Relationship
Most superficial of the three lateral abdominal muscles

Note: Important in forced expiration, coughing, sneezing, and trunk stability.

INTERNAL OBLIQUE *(Obliquus Internus Abdominis)*

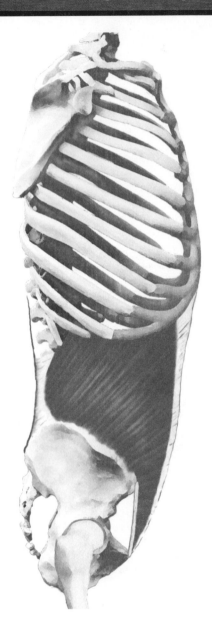

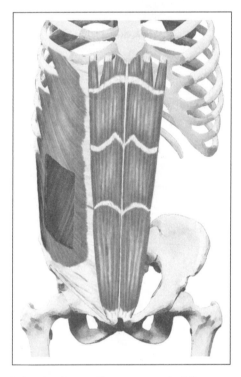

Abdominal muscles—external oblique cut

Trunk—lateral view

■ Origin
Lateral half of inguinal ligament, iliac crest, thoracolumbar fascia

■ Insertion
Cartilage of bottom three or four ribs, abdominal aponeurosis (rectus sheath) to linea alba

■ Action
Compresses abdominal contents, laterally flexes and rotates vertebral column

■ Nerve
Eighth to twelfth intercostal, iliohypogastric, ilioinguinal nerves

■ Relationship
Middle layer of the three lateral abdominal muscles

Note: Important in forced expiration, coughing, sneezing, and trunk stability.

CREMASTER

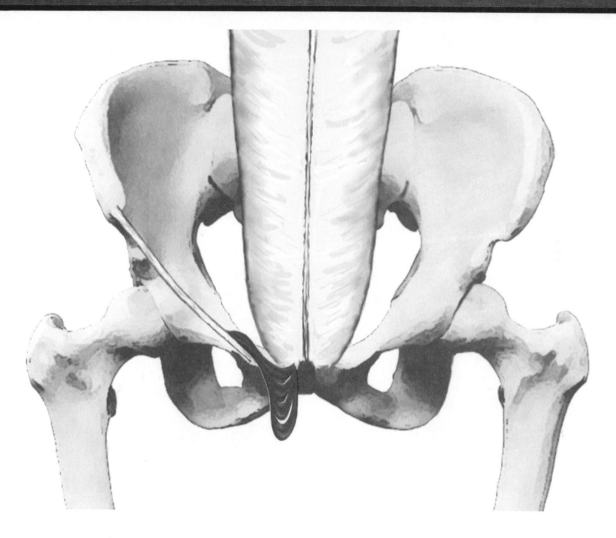

Trunk—anterior view

■ **Origin**
Inguinal ligament

■ **Insertion**
Pubic tubercle, crest of pubis, sheath of rectus abdominis

■ **Action**
Pulls testes toward body

■ **Nerve**
Genital branch of genitofemoral nerve

■ **Relationship**
This muscle is continuous with the internal abdominal oblique.

Note: The cremaster regulates the temperature of the testes, important for spermatogenesis.

TRANSVERSE ABDOMINAL *(Transversus Abdominis)*

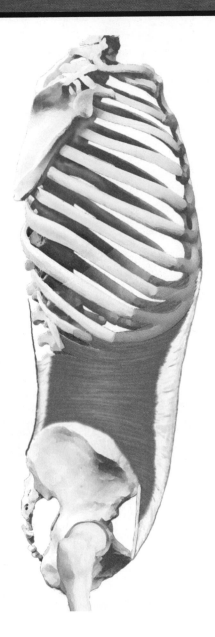

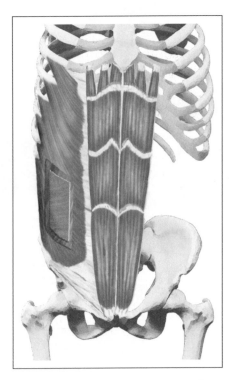

Abdominal muscles—external
and internal obliques cut

Trunk—lateral view

■ Origin
Lateral part of inguinal ligament, iliac crest,
thoracolumbar fascia, cartilage of lower six ribs

■ Insertion
Abdominal aponeurosis (rectus sheath) to linea alba

■ Action
Compresses abdomen

■ Nerve
Seventh to twelfth intercostal, iliohypogastric,
ilioinguinal nerves

■ Relationship
Deepest of the three lateral abdominal muscles

Note: Important in forced expiration, coughing, sneezing,
and trunk stability.

RECTUS ABDOMINIS*

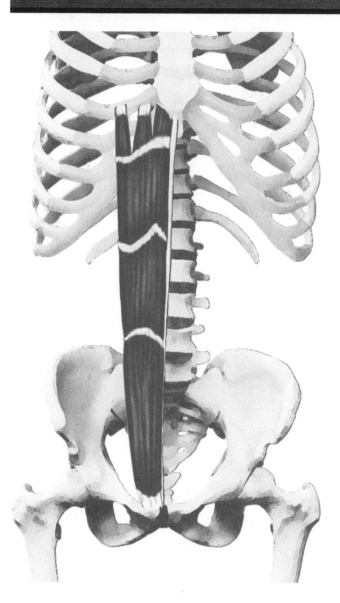

Trunk—anterior view

■ Origin
Crest of pubis, pubic symphysis

■ Insertion
Cartilage of fifth, sixth, and seventh ribs, xiphoid process

■ Action
Flexes vertebral column, compresses abdomen

■ Nerve
Seventh through twelfth intercostal nerves

*Tendinous bands divide each rectus into three or four bellies. each rectus is sheathed in aponeurotic fibers from the lateral abdominal muscles. These fibers meet centrally to form the linea alba.

Note: The pyramidalis is a small, unimportant muscle that extends from the ventral surface of the pubis to the lower part of the linea alba. It is frequently absent.

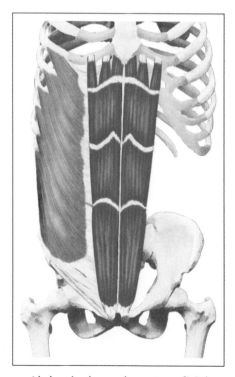

Abdominal muscles—superficial

QUADRATUS LUMBORUM

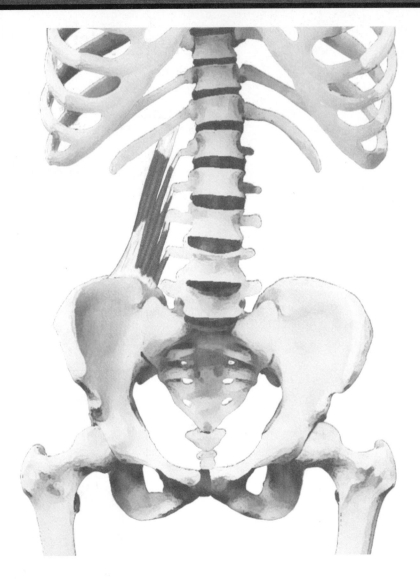

Lower trunk—anterior view

■ Origin
Iliolumbar ligament, iliac crest

■ Insertion
Twelfth rib, transverse processes of upper four lumbar vertebrae

■ Action
Laterally flexes vertebral column, fixes ribs for forced expiration

■ Nerve
T12, L1

Note: Fixation of the ribs may provide a stable attachment of the diaphragm for voice control in singers.

Muscles of the Shoulder and Arm

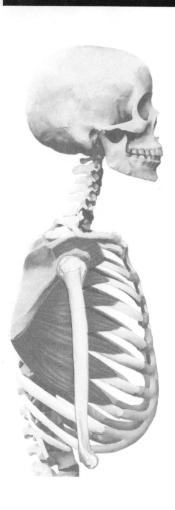

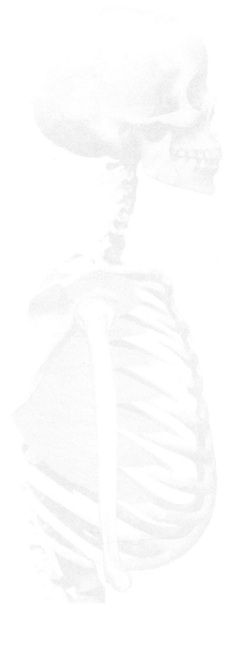

PECTORALIS MAJOR

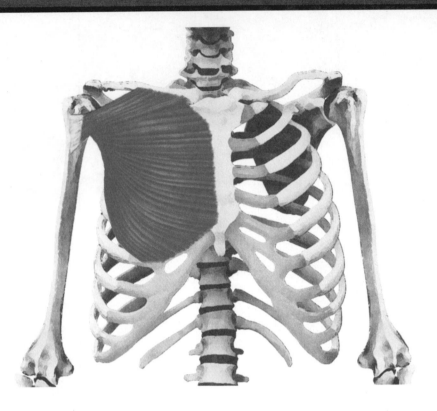

Anterior view

■ Origin
Clavicular part—medial half of the clavicle

Sternocostal part—sternum, upper six costal cartilages, aponeurosis of external oblique

■ Insertion
Lateral lip of intertubercular (bicipital) groove of humerus, crest below greater tubercle of the humerus

■ Action
Both parts adduct, medially rotate arm; clavicular part flexes arm from full extension; sternocostal part extends the flexed arm

■ Nerve
Medial and lateral pectoral nerves (C5–C8, T1)

PECTORALIS MINOR

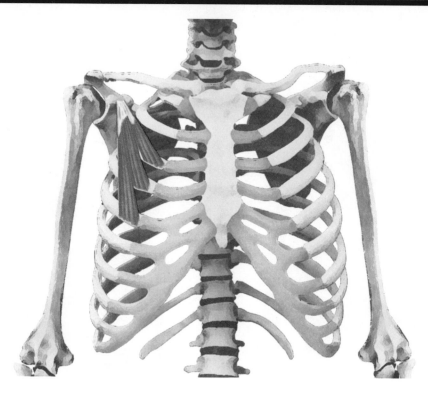

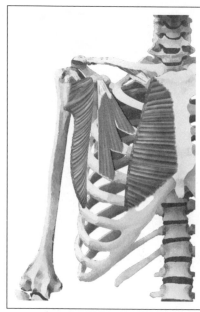

Muscles of anterior chest
(pectoralis major turned back)

Anterior view

■ Origin
External surfaces of the third, fourth, and fifth ribs

■ Insertion
Coracoid process of the scapula

■ Action
Draws scapula forward and downward, raises ribs* in forced inspiration

■ Nerve
Medial pectoral nerve (C8, T1)

■ Relationship
Deep to pectoralis major; medial pectoral nerve pierces this muscle

*Raising the ribs requires stabilization of the scapula by the rhomboids and trapezius.

SUBCLAVIUS

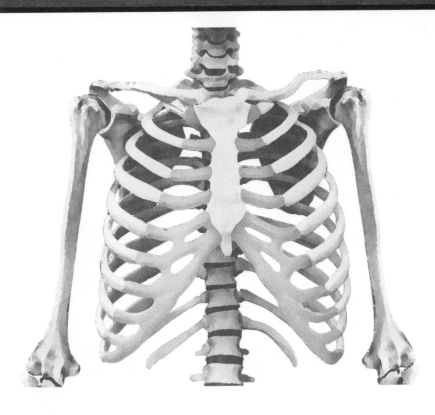

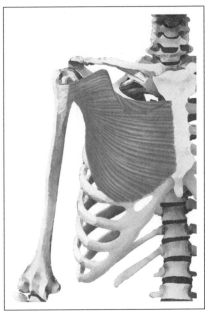

Muscles of anterior chest
(pectoralis major turned back)

Anterior view

■ Origin
Junction of the first rib with its costal cartilage

■ Insertion
Groove on the inferior (lower) surface of the clavicle

■ Action
Steadies clavicle during movements of the shoulder girdle

■ Nerve
Nerve to the subclavius from upper trunk of brachial plexus (C5, C6)

CORACOBRACHIALIS

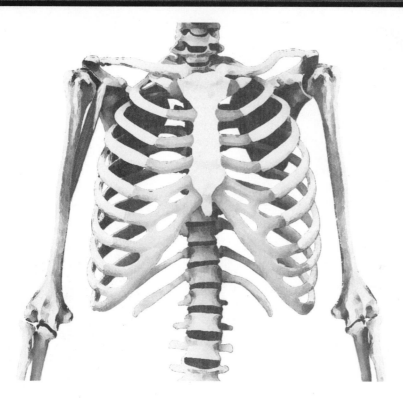

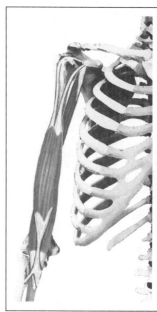

Muscles of anterior arm

Anterior view

■ **Origin**
Tip (apex) of the coracoid process of scapula

■ **Insertion**
Middle third of the medial surface and border of the humerus

■ **Action**
Weakly adducts arm (flexion unsubstantiated), aids in stabilizing humerus

■ **Nerve**
Musculocutaneous nerve (C6, C7)

■ **Relationship**
Deep to short head of biceps

Note: The musculocutaneous nerve usually pierces this muscle.

BICEPS BRACHII

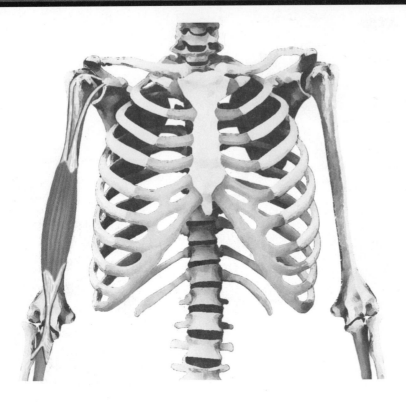

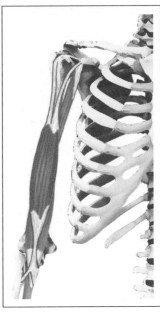

Muscles of anterior arm

Anterior view

■ Origin
Long head—supraglenoid tubercle of scapula

Short head—coracoid process of scapula

■ Insertion
Tuberosity of radius, bicipital aponeurosis into deep fascia on medial part of forearm

■ Action
Supinates forearm, flexes forearm, weakly flexes arm at shoulder

■ Nerve
Musculocutaneous nerve (C5, C6)

■ Relationship
Long head passes through intertubercular (bicipital) groove, then inside glenohumeral joint capsule

Note: As a two joint muscle, its contribution to shoulder flexion mainly occurs when both joints are extended.

BRACHIALIS

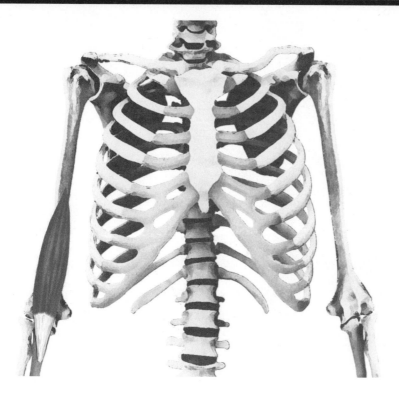

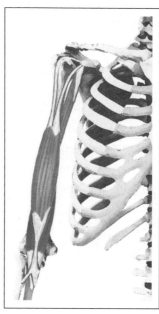

Muscles of anterior arm

Anterior view

■ **Origin**
Anterior of lower half of humerus

■ **Insertion**
Coronoid process of ulna, tuberosity of ulna

■ **Action**
Flexes forearm

■ **Nerve**
Musculocutaneous nerve (C5, C6)

■ **Relationship**
Deep to biceps brachii

MUSCLES OF THE ANTERIOR ARM AND CHEST

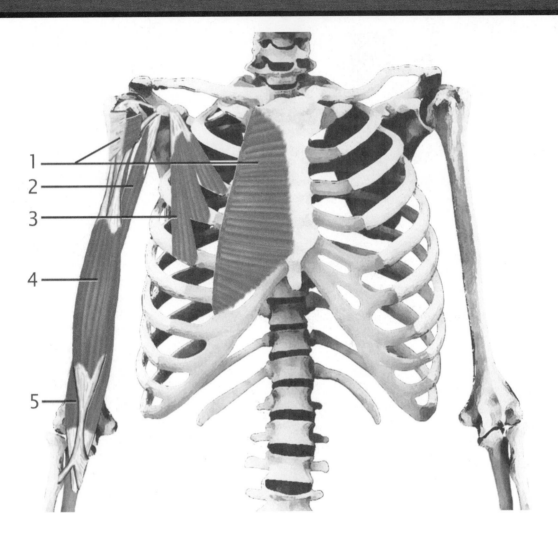

Shoulder—anterior view

1. Pectoralis major
2. Coracobrachialis
3. Pectoralis minor
4. Biceps brachii
5. Brachialis

TRAPEZIUS

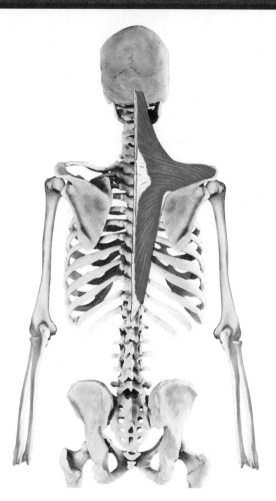

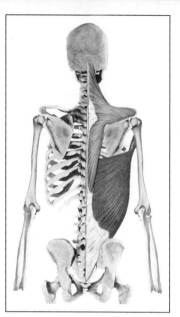

Posterior back (superficial layer)

Posterior view

■ Origin
Medial third of superior nuchal line, external occipital protuberance, ligamentum nuchae, spinous processes and supraspinous ligaments of seventh cervical and all thoracic vertebrae

■ Insertion
Upper part—lateral third of clavicle

Middle part—acromion and spine of scapula

Lower part—medial portion of spine of scapula (tubercle)

■ Action
Upper part elevates scapula*, middle part retracts (adducts) scapula, lower part depresses scapula, upper and lower parts together rotate scapula (important in elevating arm)

■ Nerve
Accessory (eleventh cranial), C3, C4

■ Relationship
Most superficial muscle of back

*Upper part stabilizes scapula against downward rotation, as when weight is carried in the hand.

LATISSIMUS DORSI

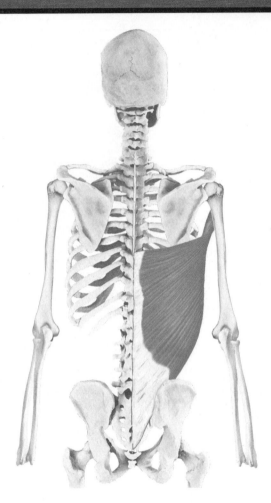

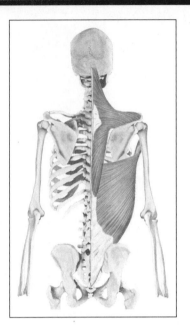

Posterior back (superficial layer)

Posterior view

■ Origin
Spinous processes of lower six thoracic vertebrae, spinous processes of all lumbar and sacral vertebrae (through thoracolumbar fascia), supraspinal ligament and iliac crest, outer surfaces of lower three or four ribs, inferior angle of scapula

■ Insertion
Bottom of intertubercular (bicipital) groove

■ Action
Extends, adducts, and medially rotates the arm; draws the shoulder downward and backward; keeps inferior angle of scapula against the chest wall; accessory muscle of respiration

■ Nerve
Thoracodorsal nerve (C6–C8)

Note: This muscle is used for the crawl (freestyle) stroke in swimming.

LEVATOR SCAPULAE

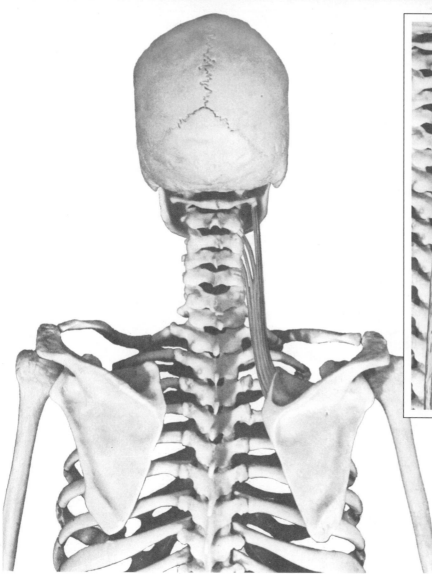

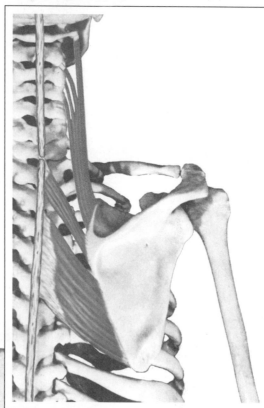

Posterior back (deep layer)

Posterior view

■ Origin
Posterior tubercles of the transverse processes of the first four cervical vertebrae

■ Insertion
Vertebral (medial) border of the scapula at and above the spine

■ Action
Elevates medial border of scapula, rotates scapula to lower the lateral angle, acts with trapezius and rhomboids to pull scapula medially and upward, bends neck laterally

■ Nerve
Dorsal scapular nerve (C5)

RHOMBOID MAJOR

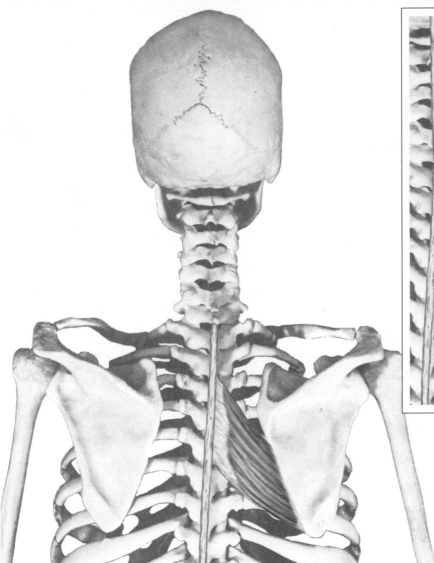

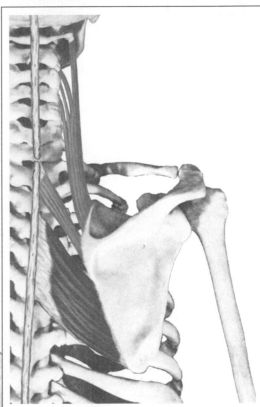

Posterior back (deep layer)

Posterior view

■ Origin
Spines of the second to fifth thoracic vertebrae, supraspinous ligament

■ Insertion
Medial border of the scapula below the spine

■ Action
Retracts and stabilizes scapula, elevates the medial border of the scapula causing downward rotation, assists in adduction of arm

■ Nerve
Dorsal scapular nerve (C5)

RHOMBOID MINOR

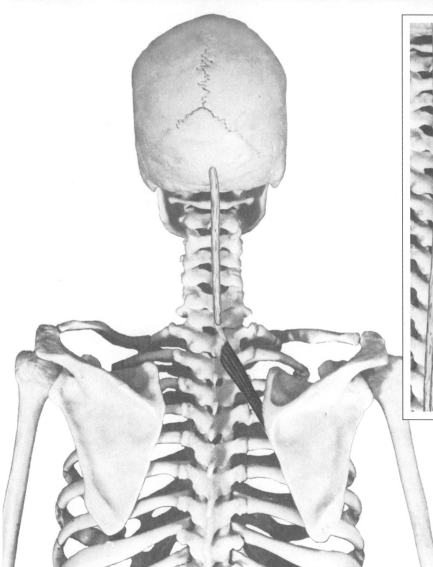

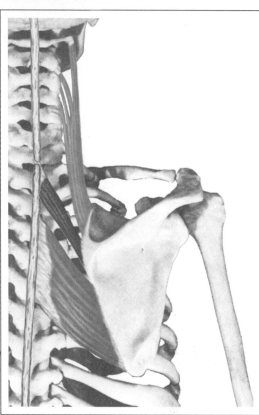

Posterior back (deep layer)

Posterior view

■ Origin
Spines of the seventh cervical and first thoracic vertebrae, lower part of the ligamentum nuchae

■ Insertion
Medial border of the scapula at the root of the spine

■ Action
Retracts and stabilizes scapula, elevates the medial border of the scapula, rotates the scapula to depress the lateral angle (assists in adduction of arm)

■ Nerve
Dorsal scapular nerve (C5)

SERRATUS ANTERIOR

Lateral view

■ Origin
Outer surfaces and superior borders of first eight or nine ribs, and fascia covering first intercostal space

■ Insertion
Anterior surface (costal surface) of the medial border of the scapula

■ Action
Rotates scapula (upward rotation) for abduction and flexion of arm, protracts scapula

■ Nerve
Long thoracic nerve (C5–C7)

■ Relationships
Serratus anterior and rhomboids both insert on the medial border of scapula; they are antagonists causing protraction and retraction; long thoracic nerve lies on the superficial surface of this muscle

Note: When the arm is fixed, this muscle can assist in rib movement for ventilation.

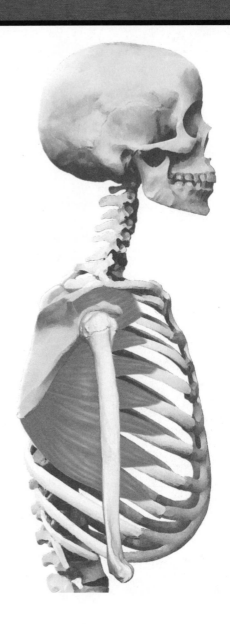

DELTOID

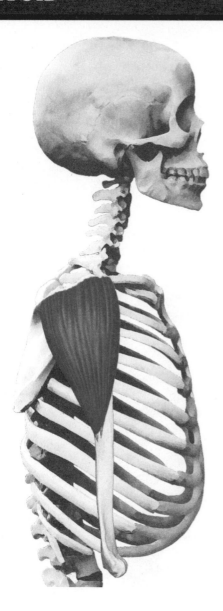

Lateral view

■ Origin

Anterior portion—anterior border and superior surface of the lateral third of the clavicle

Middle portion—lateral border of the acromion process

Posterior portion—lower border of the crest of the spine of the scapula

■ Insertion

Deltoid tuberosity, on the middle of the lateral surface of the shaft of the humerus

■ Action

Anterior portion—flexes and medially rotates arm

Middle portion—abducts arm

Posterior portion—extends and laterally rotates arm

■ Nerve

Axillary nerve (C5, C6)

SUPRASPINATUS *(Rotator Cuff*)*

Lateral view

■ Origin
Supraspinous fossa of scapula

■ Insertion
Upper part of the greater tuberosity of the humerus, capsule of the shoulder joint

■ Action
Aids deltoid in abduction of arm; draws humerus toward glenoid fossa, preventing deltoid from forcing humerus up against acromion

■ Nerve
Suprascapular nerve (C5)

*Supraspinatus, infraspinatus, teres minor, and subscapularis together are called the rotator cuff.

Note: The joint capsule and its ligaments cannot provide necessary support because of the great range of motion between the humerus and scapula. The rotator cuff muscles prevent dislocation of the humerus throughout most of the arm's range of motion. This muscle is commonly injured by repeated overhead motions (impingement syndrome).

Rotator cuff
(posterior view)

INFRASPINATUS *(Rotator Cuff)*

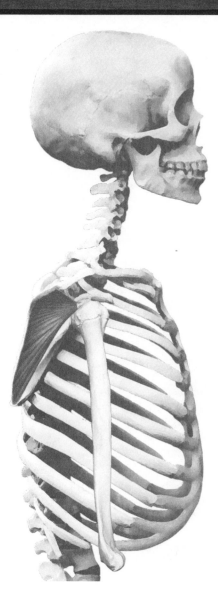

Lateral view

■ Origin
Infraspinous fossa of the scapula

■ Insertion
Middle facet of the greater tuberosity of the humerus, capsule of the shoulder joint

■ Action
Draws humerus toward glenoid fossa, thus resisting posterior dislocation of arm, as in crawling; laterally rotates; aids in stabilization of humerus during abduction

■ Nerve
Suprascapular nerve (C5, C6)

Rotator cuff
(posterior view)

TERES MINOR *(Rotator Cuff)*

Lateral view

■ Origin
Upper two-thirds of the dorsal surface of the axillary border of the scapula

■ Insertion
The capsule of the shoulder joint, the lower facet of the greater tuberosity of the humerus

■ Action
Laterally rotates arm, weakly adducts arm, draws humerus toward glenoid fossa

■ Nerve
Axillary nerve (C5)

Rotator cuff
(posterior view)

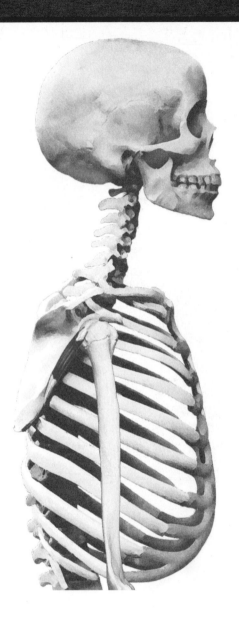

SUBSCAPULARIS *(Rotator Cuff)*

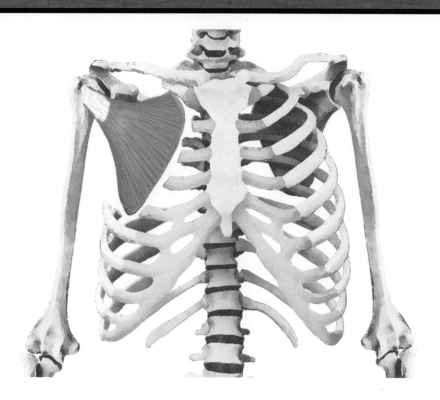

Anterior view
(Upper ribs cut away)

■ Origin
Subscapular fossa on the anterior surface of scapula

■ Insertion
Lesser tuberosity (tubercle) of the humerus, ventral part of the capsule of the shoulder joint

■ Action
Medially rotates arm, stabilizes glenohumeral joint

■ Nerve
Upper and lower subscapular nerves (C5, C6)

TERES MAJOR

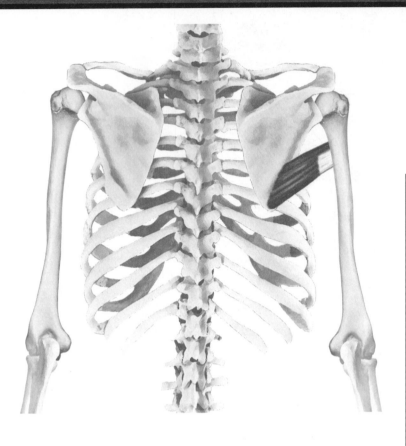

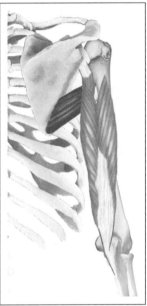

Posterior muscles of arm

Posterior view

■ Origin
Lower third of the posterior surface of the lateral border of the scapula, near the inferior angle

■ Insertion
Medial lip of the intertubercular (bicipital) groove of the humerus

■ Action
Medially rotates arm, adducts arm, extends arm

■ Nerve
Lower subscapular nerve (C5, C6)

TRICEPS BRACHII

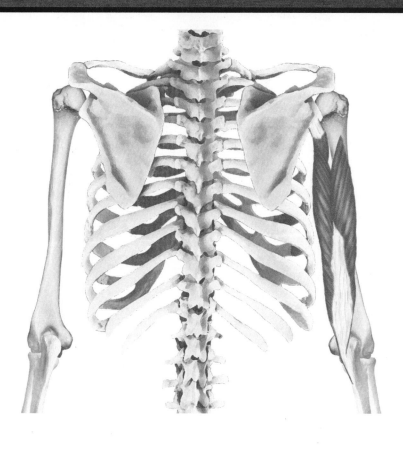

Posterior muscles of arm

Posterior view

■ Origin

Long head—infraglenoid tubercle of the scapula

Lateral head—upper half of the posterior surface of the shaft of the humerus

Medial head—posterior surface of the lower half of the shaft of the humerus

■ Insertion

Posterior part of olecranon process of the ulna

■ Action

Extends forearm, long head aids in adduction if arm is abducted

■ Nerve

Radial nerve (C7, C8)

Note: The radial nerve comes from the axilla (armpit) and passes along the humerus between the medial and lateral heads. It can be compressed against the humerus, so it is one of the most commonly injured peripheral nerves.

ANCONEUS

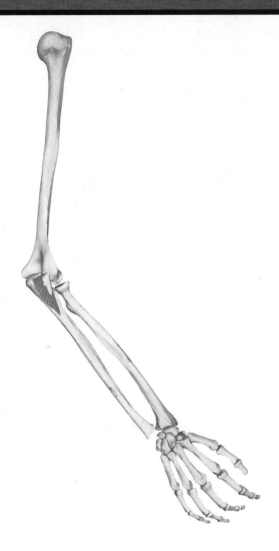

Posterior view of arm

■ Origin
Posterior part of lateral epicondyle of the humerus

■ Insertion
Lateral surface of the olecranon process and posterior surface of ulna

■ Action
Extends forearm (assists triceps)

■ Nerve
Radial nerve (C7, C8)

CHAPTER SEVEN

Muscles of the Forearm and Hand

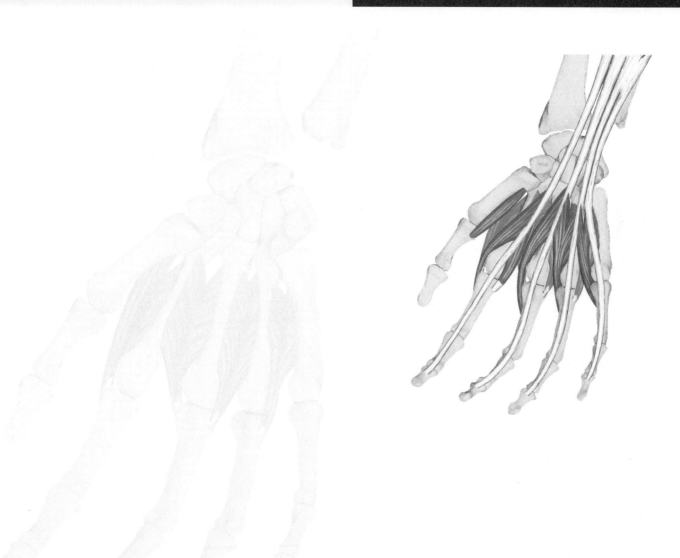

PRONATOR TERES *(Superficial Group)*

Forearm—anterior view

■ Origin
Humeral head—medial supracondylar ridge and medial epicondyle of the humerus

Ulnar head—medial border of the coronoid process of the ulna

■ Insertion
Middle of lateral surface of the radius (pronator tuberosity)

■ Action
Pronates and flexes forearm

■ Nerve
Median nerve (C6, C7)

Note: The median nerve passes between the humeral and ulnar heads of this muscle.

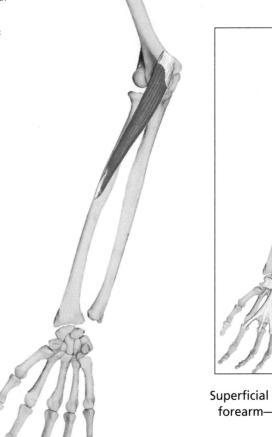

Superficial muscles of the forearm—ventral view

FLEXOR CARPI RADIALIS *(Superficial Group)*

Forearm—anterior view

■ Origin
Medial epicondyle of the humerus through the common tendon

■ Insertion
Front of the bases of the second and third metacarpal bones

■ Action
Flexes hand, synergist in abduction with extensor carpi radialis longus and brevis

■ Nerve
Median nerve (C6, C7)

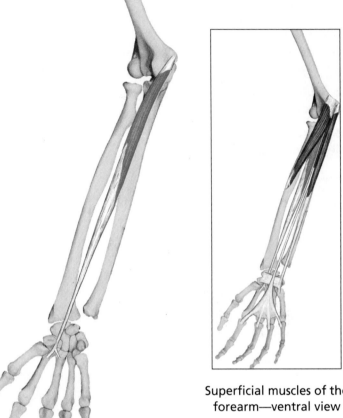

Superficial muscles of the forearm—ventral view

PALMARIS LONGUS *(Superficial Group)*

Forearm—anterior view

■ Origin
Medial epicondyle of the humerus through the common tendon

■ Insertion
Front (central part) of the flexor retinaculum and apex of the palmar aponeurosis

■ Action
Flexes the hand

■ Nerve
Median nerve (C6, C7)

Note: This muscle is absent in about 14% of limbs.

Reference: Agur, Amr: *Grant's Atlas of Anatomy*, 9th ed. Williams & Wilkins, Baltimore, 1991.

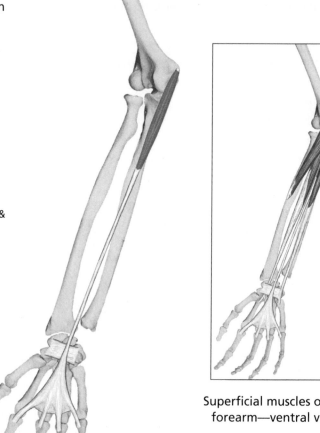

Superficial muscles of the forearm—ventral view

FLEXOR CARPI ULNARIS *(Superficial Group)*

Forearm—anterior view

■ Origin
Humeral head—medial epicondyle of the humerus through the common tendon

Ulnar head—medial margin of olecranon process of ulna, dorsal border of shaft of the ulna

■ Insertion
Pisiform bone, hook of the hamate, and base of the fifth metacarpal bone

■ Action
Flexes hand, synergist in adduction of hand with extensor carpi ulnaris

■ Nerve
Ulnar nerve (C8, T1)

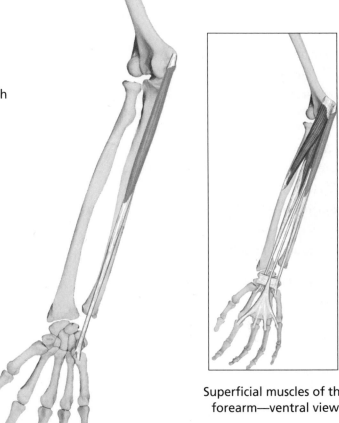

Superficial muscles of the forearm—ventral view

FLEXOR DIGITORUM SUPERFICIALIS

Forearm—anterior view

■ Origin
Humeroulnar head—medial epicondyle of the humerus through common tendon,* medial margin of the coronoid process of ulna

Radial head—anterior surface of shaft of radius

■ Insertion
Four tendons divide into two slips each; slips insert into the sides (margins of the anterior surfaces) of the middle phalanges of four fingers

■ Action
Flexes the middle phalanges of the fingers

■ Nerve
Median nerve (C7, C8, T1)

■ Relationship
Deep to superficial flexors

*See superficial flexors.

Note: The tendons of flexor digitorum superficialis split and attach to the middle phalanx. The tendons of flexor digitorum profundus pass through this split and continue to the distal phalanx.

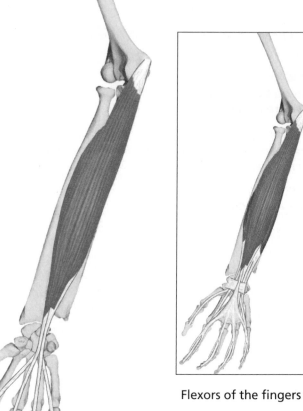

Flexors of the fingers

FLEXOR DIGITORUM PROFUNDUS

Forearm—anterior view

■ Origin
Upper three-fourths of anterior and medial surfaces of shaft of ulna and medial side of the coronoid process, interosseous membrane

■ Insertion
Front of base of distal phalanges

■ Action
Flexes distal phalanges

■ Nerve
Ulnar nerve supplies the medial half of the muscle (going to the fourth and fifth fingers)

Anterior interosseous branch of median nerve supplies lateral half (going to index and middle fingers) (C8, T1)

■ Relationship
Deep to flexor digitorum superficialis

Note: Flexor digitorum muscles, flexor pollicis longus, and the median nerve pass under the flexor retinaculum (p. 134) in the wrist. When irritated, the synovial sheaths of these muscles can compress the median nerve, causing the sensory and motor deficits known as carpal tunnel syndrome.

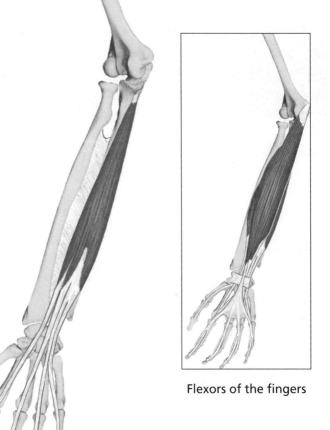

Flexors of the fingers

FLEXOR POLLICIS LONGUS

Forearm—anterior view

■ Origin
Middle of anterior surface of shaft of radius, interosseous membrane, medial epicondyle of humerus, and often coronoid process of ulna

■ Insertion
Palmar aspect of base of the distal phalanx of thumb

■ Action
Flexes the thumb

■ Nerve
Anterior interosseous branch of median nerve (C8, T1)

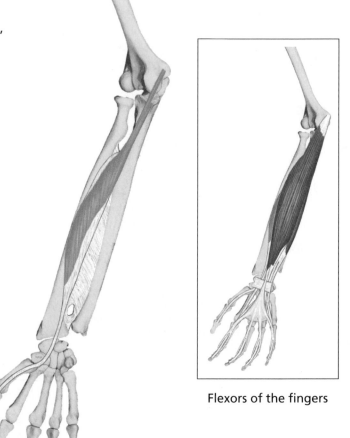

Flexors of the fingers

PRONATOR QUADRATUS

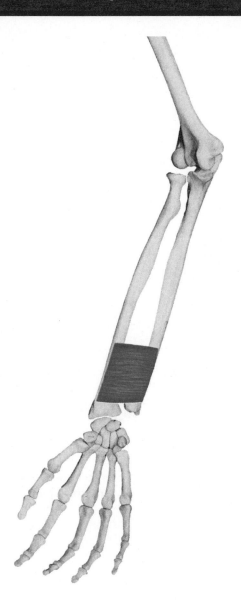

Forearm—anterior view

■ Origin
Anterior surface of distal part of shaft of ulna

■ Insertion
Lower portion of anterior surface of shaft of radius, distal part of lateral border of radius

■ Action
Pronates forearm and hand

■ Nerve
Anterior interosseous branch of median nerve (C8, T1)

■ Relationship
Deepest forearm muscle

BRACHIORADIALIS

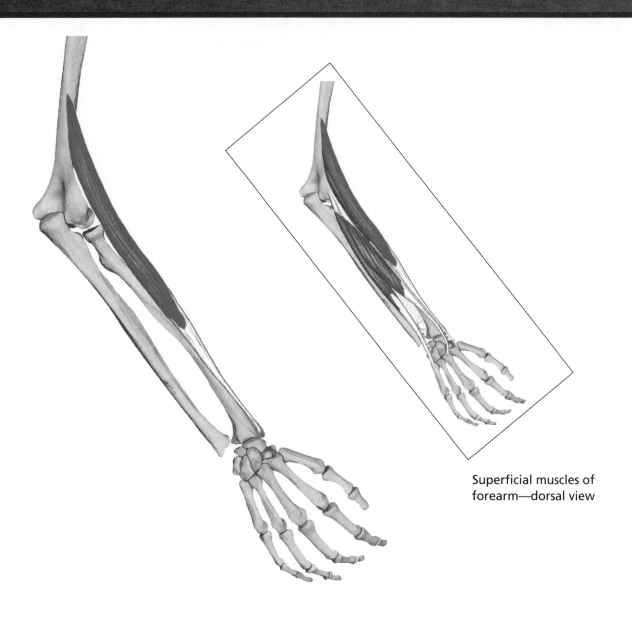

Superficial muscles of
forearm—dorsal view

Forearm—dorsal view

■ **Origin**
Upper two-thirds of lateral supracondylar ridge of
humerus

■ **Insertion**
Base of styloid process and lateral surface of radius

■ **Action**
Flexes forearm

■ **Nerve**
Radial nerve (C5, C6)

EXTENSOR CARPI RADIALIS LONGUS

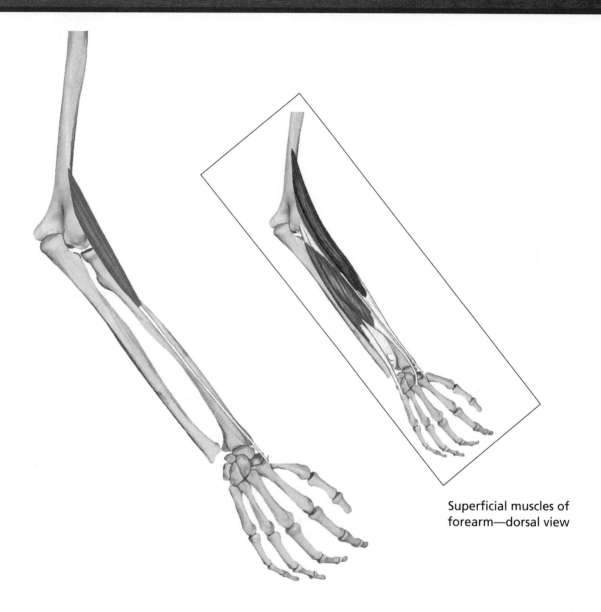

Superficial muscles of forearm—dorsal view

Forearm—dorsal view

■ Origin
Lower third of lateral supracondylar ridge of humerus

■ Insertion
Dorsal surface of the base of the second metacarpal bone

■ Action
Extends hand, synergist in abduction of hand with flexor carpi radialis

■ Nerve
Radial nerve (C6, C7)

EXTENSOR CARPI RADIALIS BREVIS

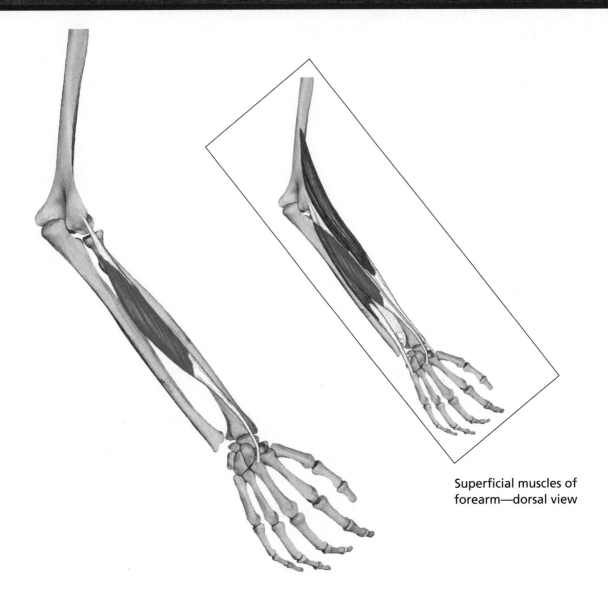

Superficial muscles of
forearm—dorsal view

Forearm—dorsal view

■ **Origin**
Lateral epicondyle of humerus

■ **Insertion**
Dorsal surface of third metacarpal bone

■ **Action**
Extends hand, synergist in abduction of hand with flexor carpi radialis

■ **Nerve**
Radial nerve (C7, C8)

EXTENSOR DIGITORUM

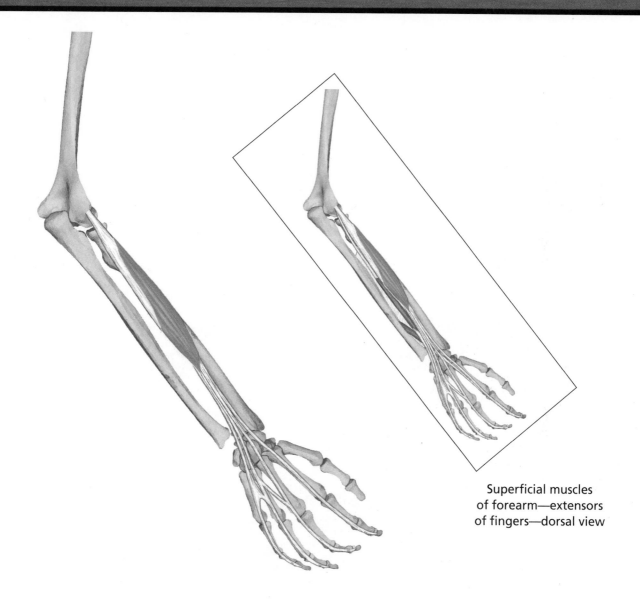

Superficial muscles
of forearm—extensors
of fingers—dorsal view

Forearm and hand—dorsal view

■ Origin
Common tendon attached to lateral epicondyle of humerus

■ Insertion
Lateral and dorsal surfaces of all the phalanges of the four fingers

■ Action
Extends the fingers and wrist

■ Nerve
Deep branch of radial nerve (C7, C8)

■ Relationship
Tends to hyperextend the metacarpophalangeal joint causing "claw hand"; its action is balanced by the lumbricales and interossei

EXTENSOR DIGITI MINIMI

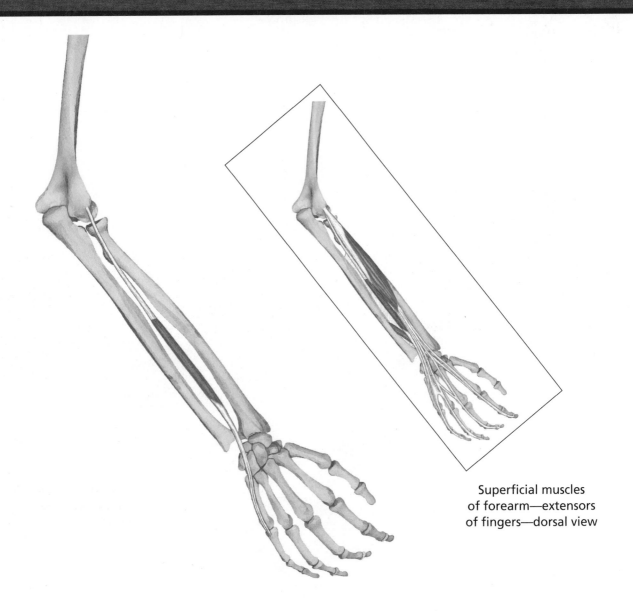

Superficial muscles
of forearm—extensors
of fingers—dorsal view

Forearm and hand—dorsal view

■ **Origin**
Common tendon attached to lateral epicondyle of
humerus

■ **Insertion**
Dorsal surface of base of first phalanx of fifth finger

■ **Action**
Extends fifth finger

■ **Nerve**
Radial nerve (C7, C8)

EXTENSOR CARPI ULNARIS

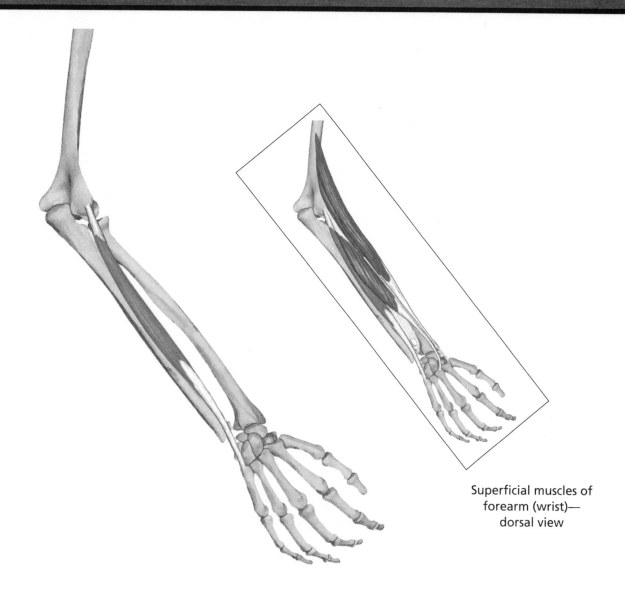

Superficial muscles of forearm (wrist)—dorsal view

Forearm and hand—dorsal view

■ **Origin**
Common tendon attached to lateral epicondyle of humerus

■ **Insertion**
Dorsal surface of base of fifth metacarpal bone

■ **Action**
Extends hand, synergist in adduction of hand with flexor carpi ulnaris

■ **Nerve**
Radial nerve (C7, C8)

SUPINATOR

Forearm and hand—anterior view

■ Origin
Lateral epicondyle of humerus, lateral ligament (radial collateral) of elbow, annular ligament of superior radioulnar joint, supinator crest of ulna

■ Insertion
Dorsal and lateral surfaces of upper third of radius

■ Action
Supinates forearm

■ Nerve
Radial nerve (C5, C6)

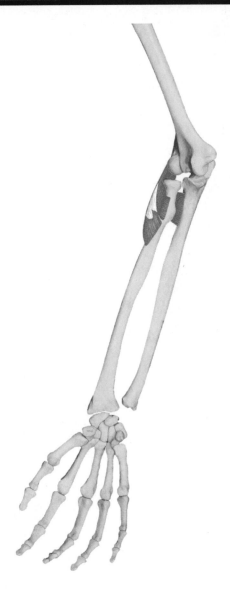

ABDUCTOR POLLICIS LONGUS

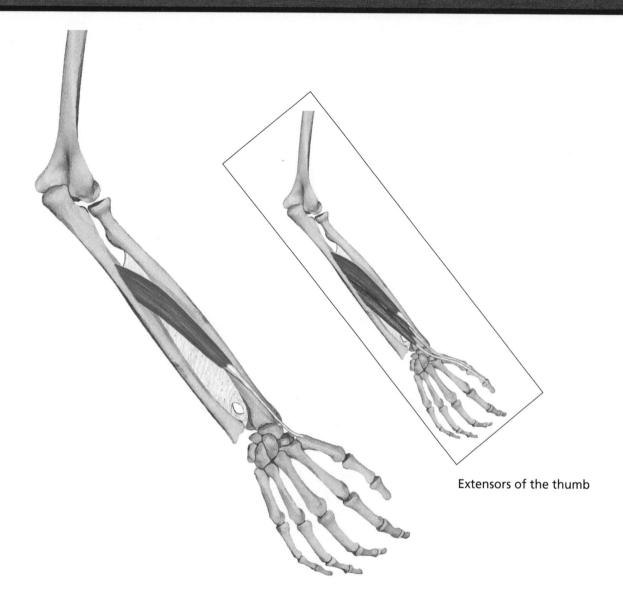

Extensors of the thumb

Forearm and hand—dorsal view

■ Origin
Posterior (dorsal) surface of shaft of radius, ulna, interosseous membrane

■ Insertion
Dorsal surface of base of first metacarpal bone

■ Action
Abducts, laterally rotates, and extends thumb; abducts wrist

■ Nerve
Radial nerve (C7, C8)

EXTENSOR POLLICIS BREVIS

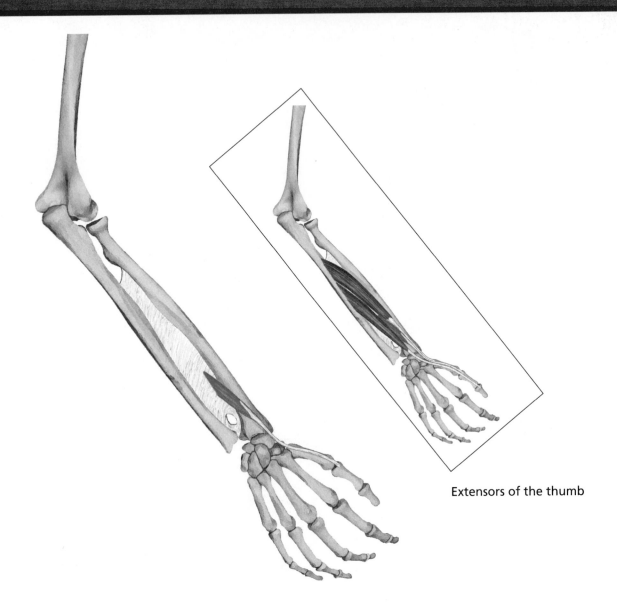

Extensors of the thumb

Forearm and hand—dorsal view

■ Origin
Dorsal surface of radius, adjacent part of interosseous membrane

■ Insertion
Base of proximal phalanx of thumb

■ Action
Extends thumb, abducts hand

■ Nerve
Radial nerve (C7, C8)

EXTENSOR POLLICIS LONGUS

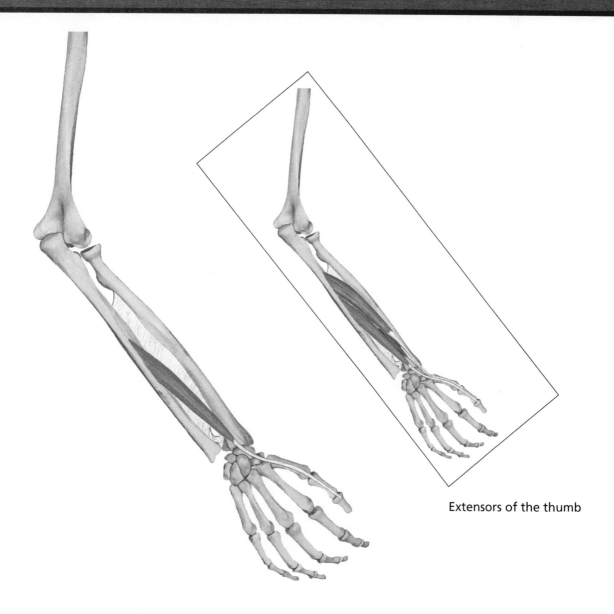

Extensors of the thumb

Forearm and hand—dorsal view

■ **Origin**
Middle third of dorsal surface of ulna, interosseous membrane

■ **Insertion**
Base of distal phalanx of thumb

■ **Action**
Extends thumb

■ **Nerve**
Radial nerve (C7, C8)

EXTENSOR INDICIS

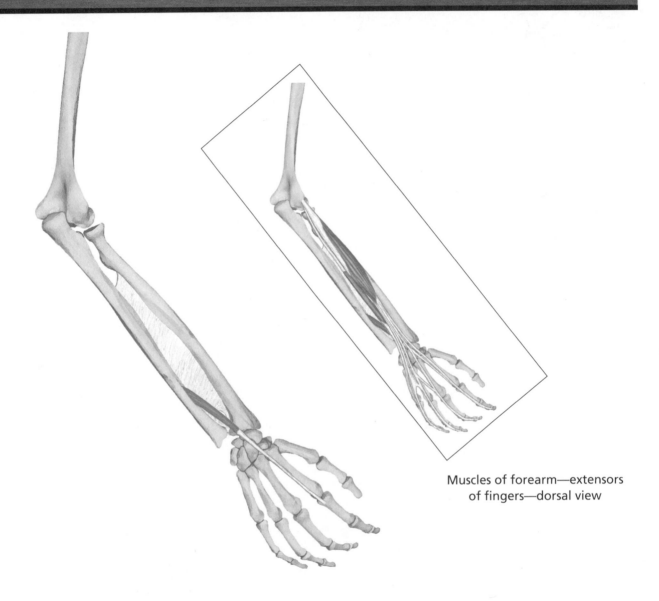

Muscles of forearm—extensors
of fingers—dorsal view

Forearm and hand—dorsal view

■ Origin
Posterior surface of ulna and adjacent part of interosseous membrane

■ Insertion
Extensor expansion on dorsal surface of proximal phalanx of index finger

■ Action
Extends index finger

■ Nerve
Radial nerve (C7, C8)

PALMARIS BREVIS

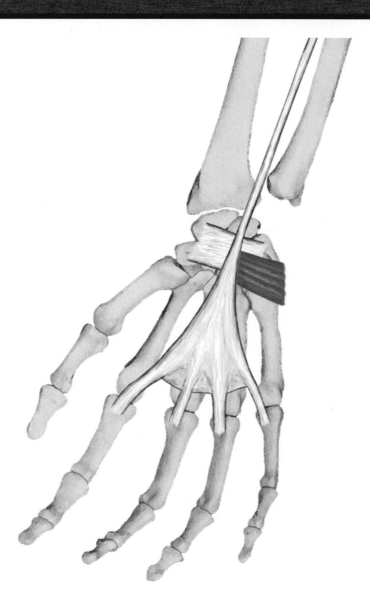

Hand—palmar view

■ Origin
Flexor retinaculum, palmar aponeurosis

■ Insertion
Skin of the palm

■ Action
Corrugates lateral skin of palm

■ Nerve
Ulnar nerve (C8, T1)

ABDUCTOR POLLICIS BREVIS *(Thenar Eminence)*

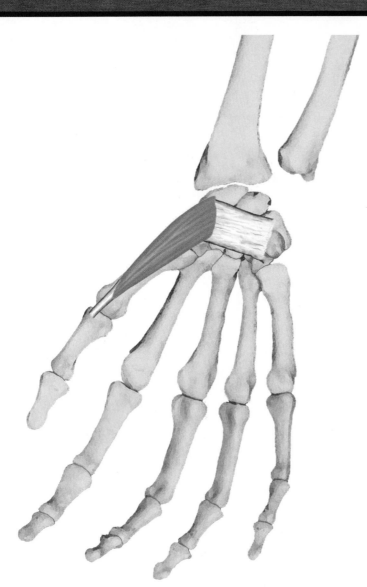

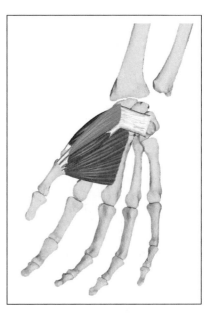

Intrinsic muscles of thumb

Hand—palmar view

■ Origin
Tubercle of scaphoid, tubercle of trapezium, flexor retinaculum

■ Insertion
Base of proximal phalanx of thumb

■ Action
Abducts thumb and moves it anteriorly, acts together with other muscles of thenar eminence to oppose thumb to other fingers

■ Nerve
Median (C8, T1)

Note: The abductor pollicis brevis, flexor pollicis brevis, and opponens pollicis form the thenar eminence at the base of the thumb.

FLEXOR POLLICIS BREVIS *(Thenar Eminence)*

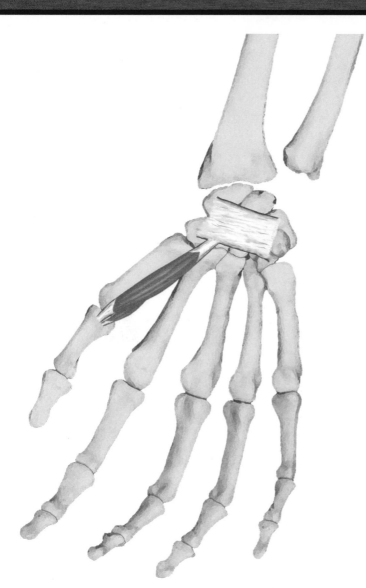

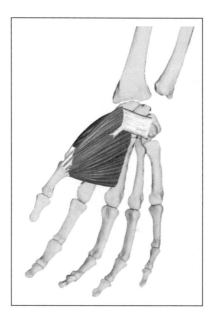

Intrinsic muscles of thumb

Hand—palmar view

■ Origin
Flexor retinaculum and trapezium, and first metacarpal bone

■ Insertion
Base of proximal phalanx of thumb

■ Action
Flexes metacarpophalangeal joint of thumb, assists in abduction and rotation of thumb, acts together with other muscles of thenar eminence to oppose thumb to other fingers

■ Nerve
Lateral portion—median nerve (C8, T1)

Medial portion—ulnar nerve (C8, T1)

OPPONENS POLLICIS *(Thenar Eminence)*

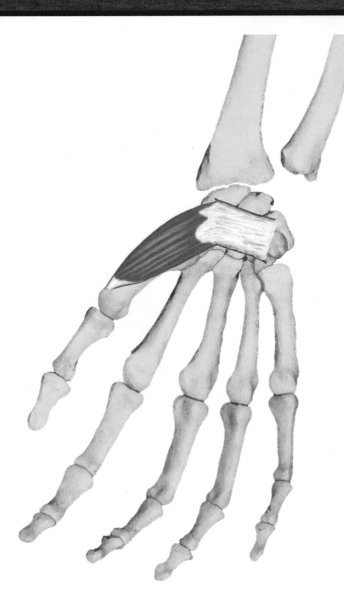

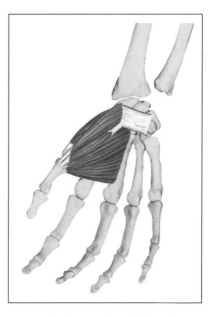

Intrinsic muscles of thumb

Hand—palmar view

■ Origin
Flexor retinaculum, tubercle of trapezium

■ Insertion
Lateral border of first metacarpal bone

■ Action
Rotates thumb into opposition with fingers, acts together with other muscles of thenar eminence to oppose thumb to other fingers

■ Nerve
Median nerve (C8, T1)

ADDUCTOR POLLICIS

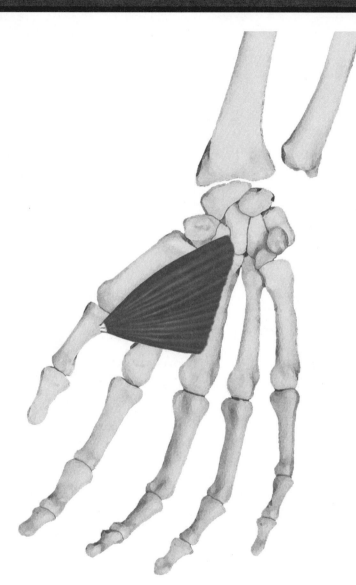

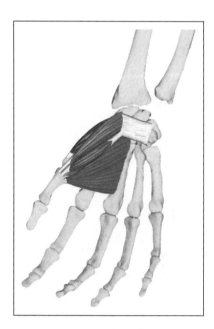

Intrinsic muscles of thumb

Hand—palmar view

■ Origin
Oblique head—anterior surfaces of second and third metacarpals, capitate, trapezoid

Transverse head—anterior surface of third metacarpal bone

■ Insertion
Medial side of base of proximal phalanx of thumb

■ Action
Adducts thumb

■ Nerve
Ulnar nerve (C8, T1)

ABDUCTOR DIGITI MINIMI *(Hypothenar Eminence)*

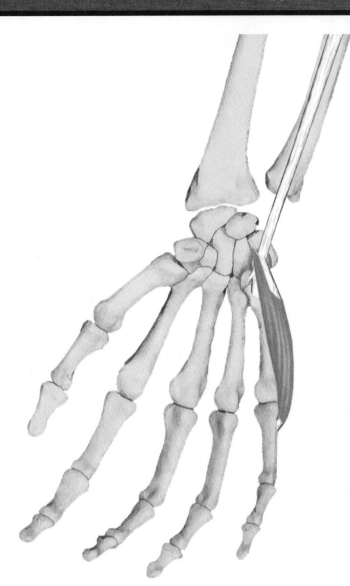

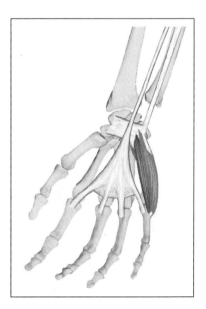

Hypothenar eminence

Hand—palmar view

■ **Origin**
Pisiform bone, tendon of flexor carpi ulnaris

■ **Insertion**
Medial side of base of proximal phalanx of fifth finger

■ **Action**
Abducts fifth finger

■ **Nerve**
Ulnar nerve (C8, T1)

Note: The hypothenar eminence is less prominent than the thenar eminence, and the fifth finger obviously cannot oppose the other digits.

FLEXOR DIGITI MINIMI BREVIS *(Hypothenar Eminence)*

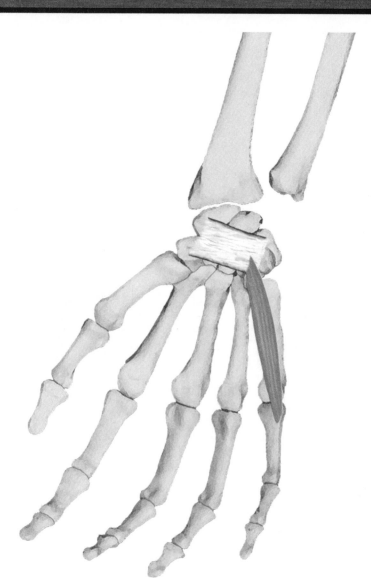

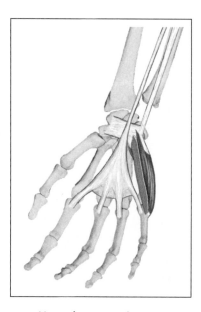

Hypothenar eminence

Hand—palmar view

■ **Origin**
Anterior surface of flexor retinaculum, hook of hamate

■ **Insertion**
Medial side of base of proximal phalanx of fifth finger

■ **Action**
Flexes fifth finger at metacarpophalangeal joint

■ **Nerve**
Ulnar nerve (C8, T1)

OPPONENS DIGITI MINIMI *(Hypothenar Eminence)*

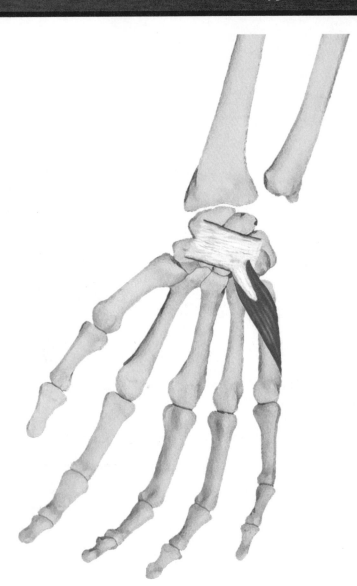

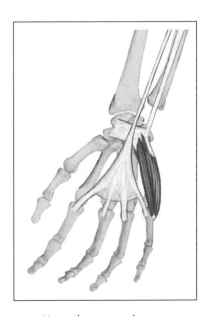

Hypothenar eminence

Hand—palmar view

■ **Origin**
Anterior surface of flexor retinaculum, hook of hamate

■ **Insertion**
Whole length of medial border of fifth metacarpal bone

■ **Action**
Rotates fifth metacarpal bone, draws fifth metacarpal bone forward, assists flexor digiti minimi brevis in flexing carpometacarpal joint of fifth finger

■ **Nerve**
Ulnar nerve (C8, T1)

LUMBRICALES* *(Four Muscles)*

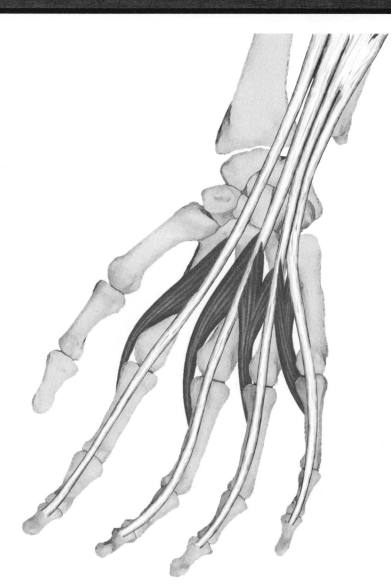

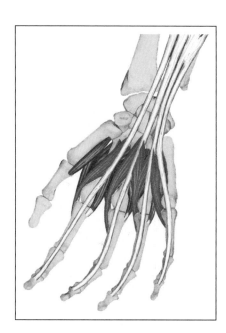

Deep muscles of hand

Hand—palmar view

■ Origin
Tendons of flexor digitorum profundus in palm

■ Insertion
Lateral side of corresponding tendon of extensor digitorum on fingers

■ Action
Extend fingers at interphalangeal joints, weakly flex fingers at metacarpophalangeal joints

■ Nerve
Lateral lumbricals (first and second)—median nerve (C8, T1)

Medial lumbricals (third and fourth)—ulnar nerve (C8, T1)

■ Relationship
Assist extensor digitorum communis in extending fingers without hyperextension at the metacarpophalangeal joints

*Associated with the tendons of flexor digitorum profundus.

PALMAR INTEROSSEI

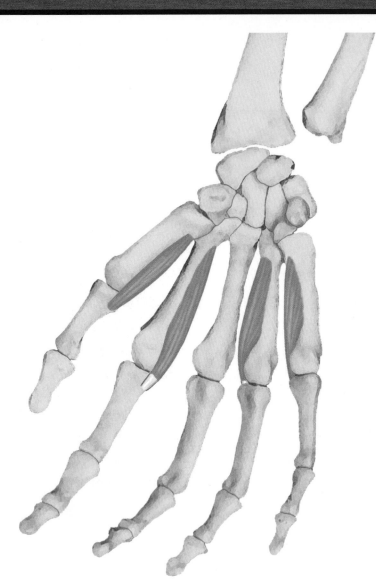

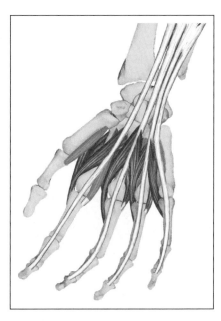

Deep muscles of hand

Hand—palmar view

■ Origin

First—medial side of base of first metacarpal bone

Second, third, and fourth—anterior surfaces of second, fourth, and fifth metacarpal bones

■ Insertion

First—medial side of base of proximal phalanx of thumb

Second—medial side of base of proximal phalanx of index finger

Third and fourth—lateral side of proximal phalanges of ring finger and fifth finger

■ Action

Adduct fingers toward center of third finger at metacarpophalangeal joints, assist in flexion of fingers at metacarpophalangeal joints

■ Nerve

Ulnar nerve (C8, T1)

Note: The palmar interosseus of the thumb is called the palmar interosseus of Henle. Most anatomists claim that it is usually absent while some argue for its common occurrence.

DORSAL INTEROSSEI

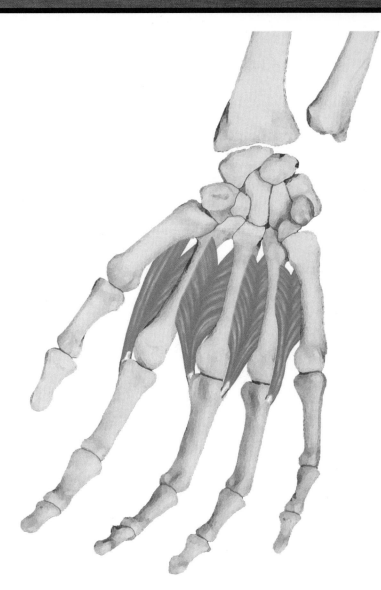

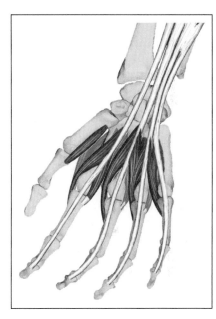

Deep muscles of hand

Hand—palmar view

■ Origin
By two heads from adjacent sides of first and second, second and third, third and fourth, and fourth and fifth metacarpal bones

■ Insertion
First—lateral side of base of proximal phalanx of index finger

Second—lateral side of base of proximal phalanx of middle finger

Third—medial side of base of proximal phalanx of middle finger

Fourth—medial side of base of proximal phalanx of ring finger

■ Action
Abduct fingers away from center of third finger at metacarpophalangeal joints, assist in flexion of fingers at metacarpophalangeal joints

■ Nerve
Ulnar nerve (C8, T1)

CHAPTER EIGHT

Muscles of the Hip and Thigh

PSOAS MAJOR *(Part of Iliopsoas)*

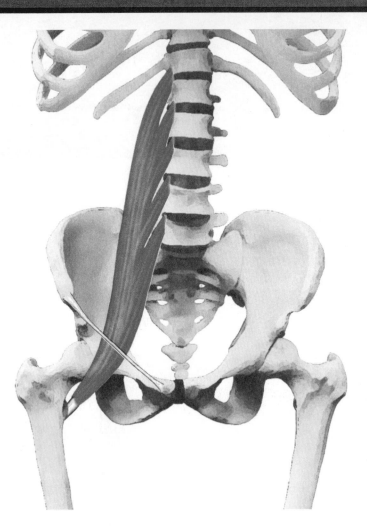

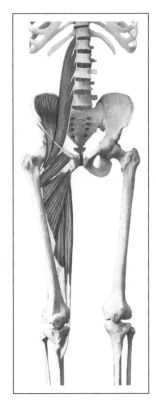

Hip flexors and adductors

Lumbar region, hip, and thigh—anterior view

■ Origin

Bases of transverse processes of all lumbar vertebrae, bodies of twelfth thoracic and all lumbar vertebrae, intervertebral disks above each lumbar vertebra

■ Insertion

Lesser trochanter of femur

■ Action

Flexes thigh,* flexes vertebral column

*See note on p. 186.

■ Nerve

Branches from lumbar plexus (L2, L3) and sometimes L1 or L4

Note: Some upper fibers insert onto the hip bone from the arcuate line to the iliopectineal eminence to form the *psoas minor*. This muscle has little function and is frequently absent.

ILIACUS *(Part of Iliopsoas)*

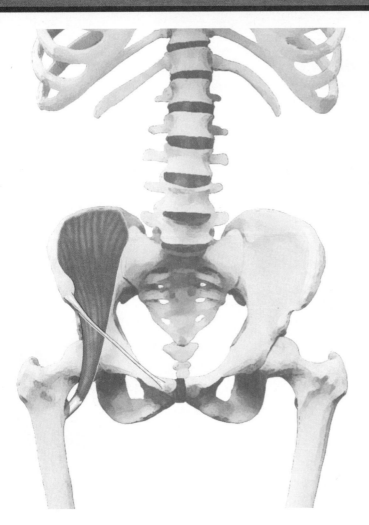

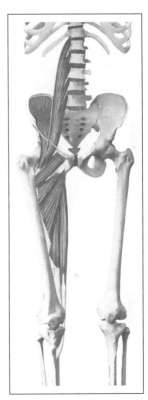

Hip flexors and adductors

Lumbar region, hip, and thigh—anterior view

■ Origin
Upper two-thirds of iliac fossa, ala of sacrum and adjacent ligaments, anterior inferior iliac spine

■ Insertion
Onto tendon of psoas major, which continues into lesser trochanter of femur (together the two muscles form the iliopsoas)

■ Action
Flexes thigh*

■ Nerve
Femoral nerve (L2, L3)

Note: The iliacus brings swinging leg forward in walking or running.

*See note on p. 186.

PIRIFORMIS

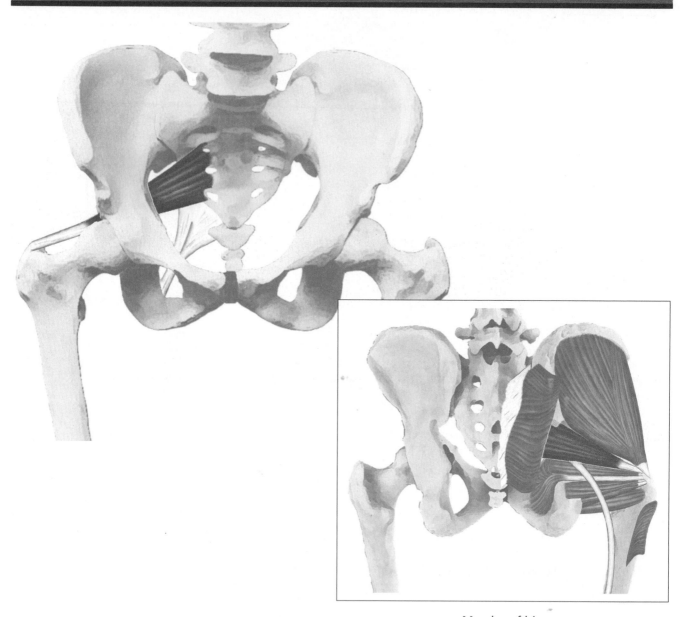

Muscles of hip

Hip and thigh—anterior view

■ Origin
Internal surface of sacrum, sacrotuberous ligament

■ Insertion
Upper border of greater trochanter of femur

■ Action
Laterally rotates thigh, abducts thigh

■ Nerve
Anterior rami of first and second sacral nerves

Note: The common peroneal part of the sciatic nerve may emerge through the belly of the piriformis instead of below its inferior border along with the tibial part. This muscle may compress the sciatic nerve (piriformis syndrome).

OBTURATOR INTERNUS

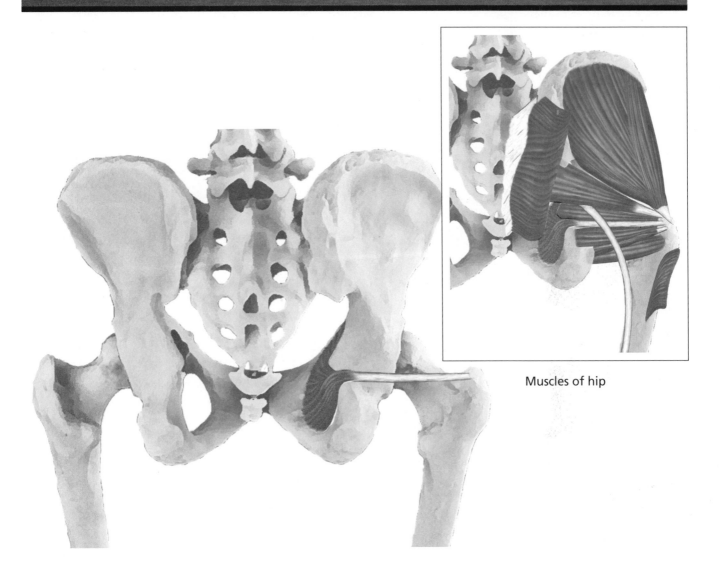

Muscles of hip

Hip—posterior view

■ Origin
Pelvic surface of obturator membrane and surrounding bones (ilium, ischium, pubis)

■ Insertion
Common tendon with superior and inferior gemelli to medial surface of upper border of greater trochanter of femur

■ Action
Laterally rotates thigh

■ Nerve
Nerve from sacral plexus (L5, S1)

GEMELLUS SUPERIOR

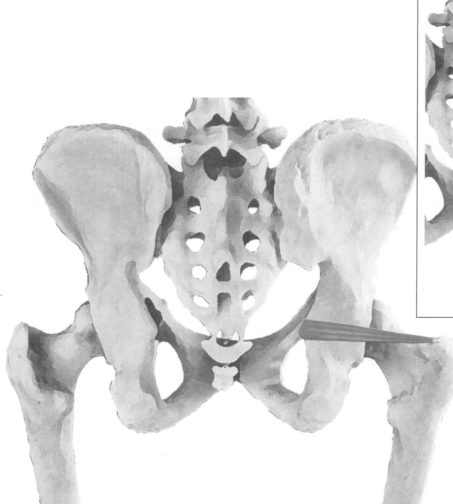

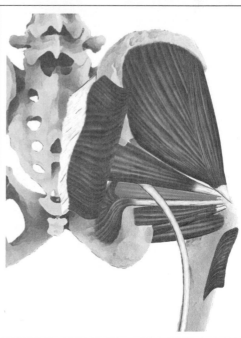

Muscles of hip

Hip—posterior view

■ Origin
Spine of ischium

■ Insertion
With tendon of obturator internus into medial surface of upper border of greater trochanter of femur

■ Action
Laterally rotates thigh

■ Nerve
Branch of nerve to obturator internus from sacral plexus (L5, S1)

GEMELLUS INFERIOR

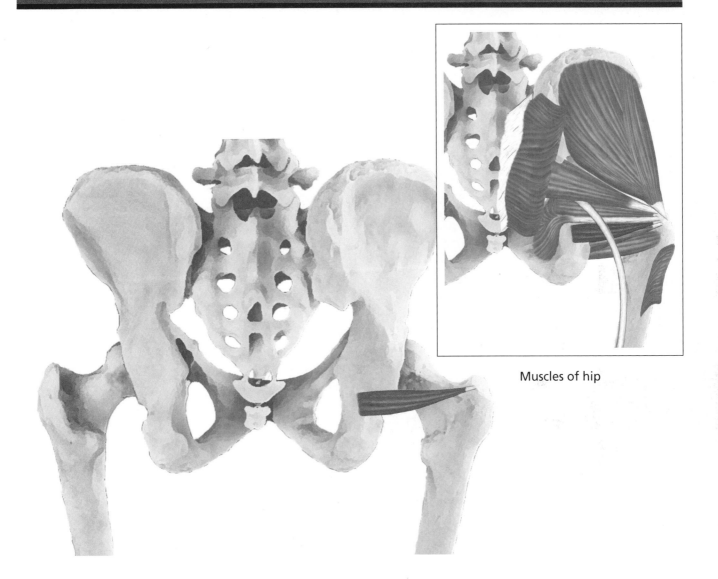

Muscles of hip

Hip—posterior view

■ Origin
Upper margin of ischial tuberosity

■ Insertion
With tendon of obturator internus into medial surface of upper border of greater trochanter of femur

■ Action
Laterally rotates thigh

■ Nerve
Branch of nerve to quadratus femoris from sacral plexus

OBTURATOR EXTERNUS

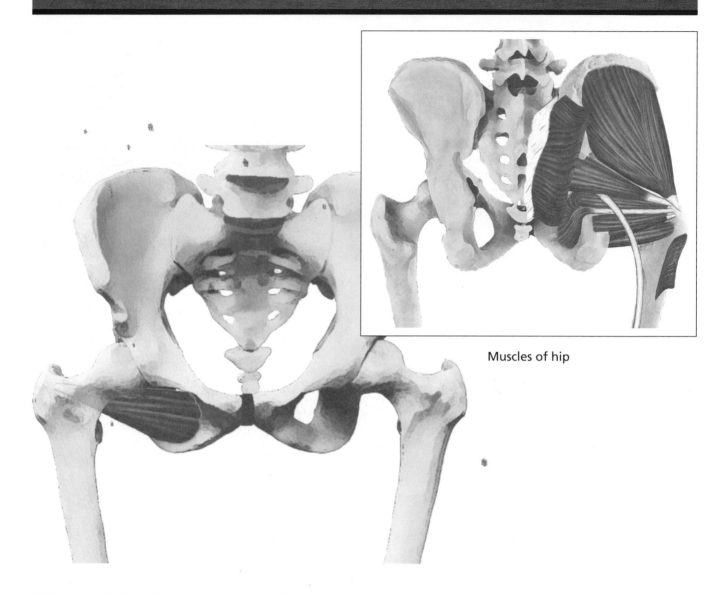

Muscles of hip

Hip and thigh—anterior view

■ Origin
Outer surface of superior and inferior rami of pubis and ramus of ischium surrounding obturator foramen

■ Insertion
Trochanteric fossa of femur

■ Action
Laterally rotates thigh

■ Nerve
Obturator nerve (L3, L4)

Note: Part of this muscle can be seen posteriorly by separating the gemellus inferior and quadratus femoris. It is deep within this cleft.

QUADRATUS FEMORIS

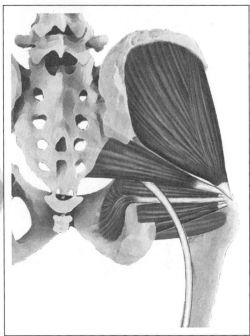

Muscles of hip

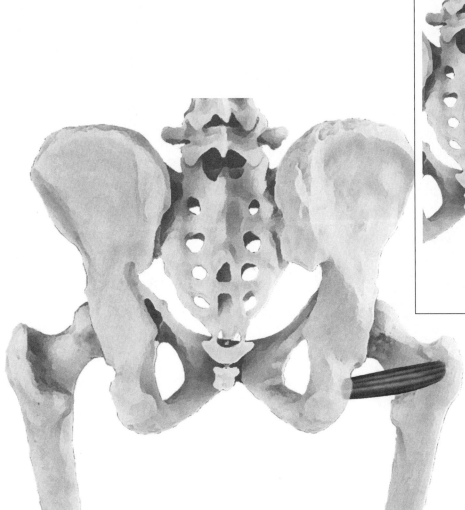

Hip and thigh—posterior view

■ **Origin**
Lateral border of ischial tuberosity

■ **Insertion**
Below intertrochanteric crest (quadrate line) of femur

■ **Action**
Laterally rotates thigh

■ **Nerve**
Branch from sacral plexus (L5, S1)

GLUTEUS MAXIMUS

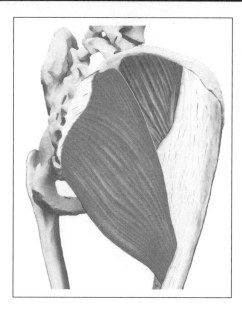

Gluteal region

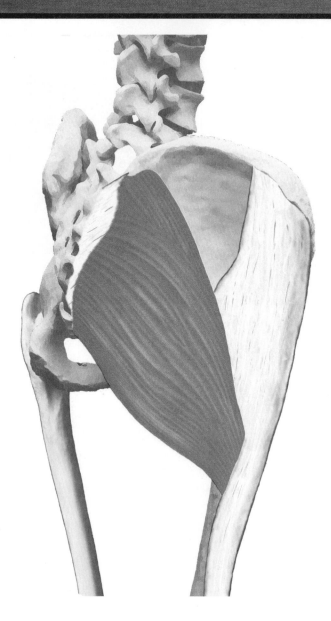

Hip and thigh—lateral view

■ Origin
Outer surface of ilium behind posterior gluteal line, adjacent posterior surface of sacrum and coccyx, sacrotuberous ligament, aponeurosis of erector spinae (sacrospinalis)

■ Insertion
Iliotibial tract of fascia lata, gluteal tuberosity of femur

■ Action
Upper part—abducts, laterally rotates thigh

Lower part—extends, laterally rotates thigh, extends trunk, assists in adduction of thigh

■ Nerve
Inferior gluteal nerve (L5, S1, S2)

Note: This is not a postural muscle; it is not used in walking but only in forceful extension, as in running, climbing, or rising from a seated position.

GLUTEUS MEDIUS

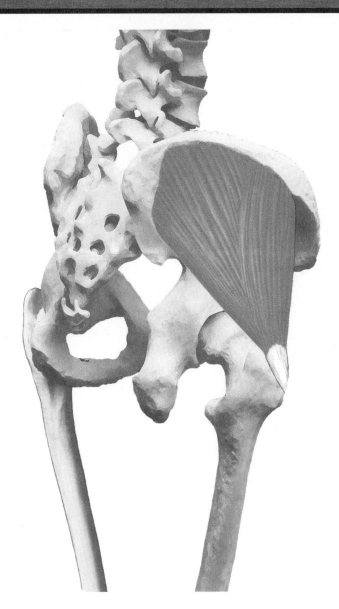

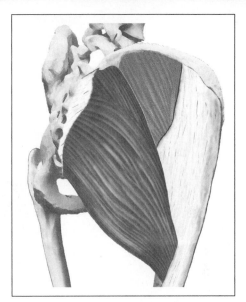

Gluteal region

Hip and thigh—lateral view

■ **Origin**
Outer surface of ilium inferior to iliac crest

■ **Insertion**
Lateral surface of greater trochanter of femur

■ **Action**
Abducts femur and rotates thigh medially

■ **Nerve**
Superior gluteal nerve (L4, L5, S1)

Note: In locomotion, this muscle (along with the gluteus minimus) prevents the pelvis from dropping (adduction of thigh) toward the opposite swinging leg.

GLUTEUS MINIMUS

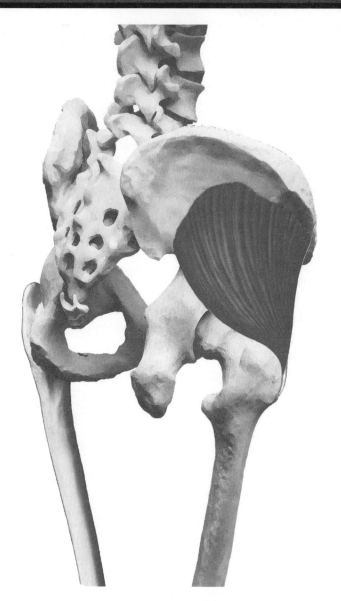

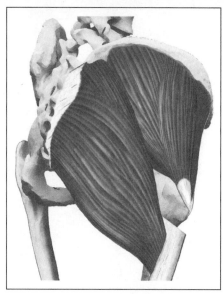

Gluteal region—iliotibial tract cut

Hip and thigh—lateral view

■ Origin
Outer surface of ilium between middle (anterior) and inferior gluteal lines

■ Insertion
Anterior surface of greater trochanter of femur

■ Action
Abducts femur, rotates thigh medially

■ Nerve
Superior gluteal nerve (L4, L5, S1)

See note on p. 173.

TENSOR FASCIAE LATAE

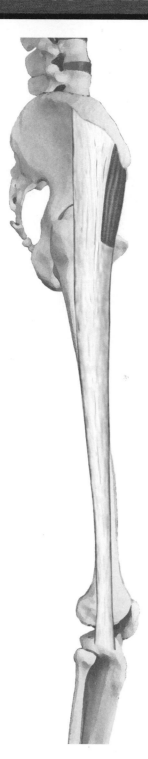

Hip and thigh—lateral view

■ Origin
Outer edge of iliac crest between anterior superior iliac spine and iliac tubercle

■ Insertion
Iliotibial tract on upper part of thigh

■ Action
Flexes, abducts thigh

■ Nerve
Superior gluteal nerve (L4, L5, S1)

Note: This muscle, along with gluteus maximus, draws the fascia lata upward, stabilizing the knee.

SARTORIUS

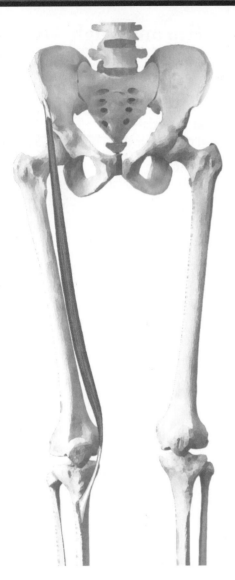

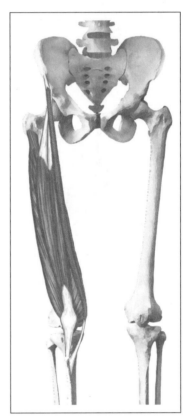

Hip flexors and adductors

Hip, thigh, and leg—anterior view

■ **Origin**
Anterior superior iliac spine and area immediately below it

■ **Insertion**
Upper part of medial surface of shaft of tibia

■ **Action**
Flexes, abducts, and laterally rotates thigh, flexes and slightly medially rotates leg at knee joint after flexion

■ **Nerve**
Femoral nerve (L2, L3)

■ **Relationship**
Insertions of sartorius, gracilis, and semitendinosus fuse on the medial tibia; these tendons, called the pes anserinus (goose foot), give medial support to the knee

Note: This muscle is used to bring swinging leg forward in walking and running.

RECTUS FEMORIS *(One of Quadriceps Femoris)*

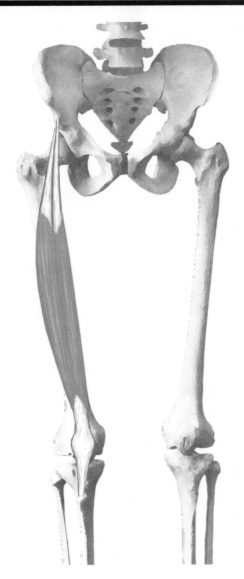

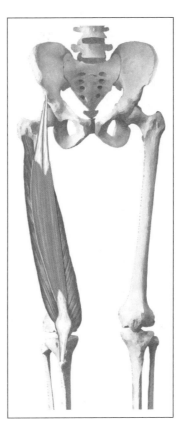

Quadriceps

Hip, thigh, and leg—anterior view

■ Origin
Anterior head—anterior inferior iliac spine

Posterior head—ilium above acetabulum

■ Insertion
Patella, then by patellar ligament to tuberosity of the tibia

■ Action
Extends leg at knee joint, flexes thigh

■ Nerve
Femoral nerve (L2–L4)

Note: This muscle is used when thigh flexion and leg extension are needed together, such as in kicking a football. In walking, the quadriceps prevent the knee from flexing during heel strike and early support phase.

VASTUS LATERALIS *(One of Quadriceps Femoris)*

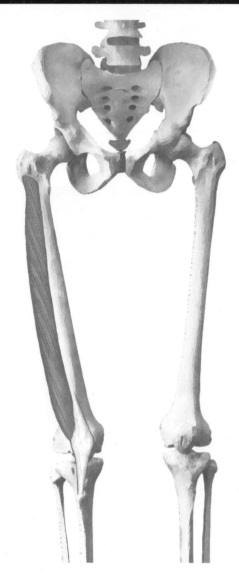

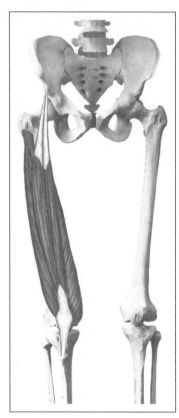

Quadriceps

Hip, thigh, and leg—anterior view

■ Origin
Intertrochanteric line, inferior border of greater trochanter, gluteal tuberosity, lateral lip of linea aspera of femur

■ Insertion
Lateral margin of patella, then by patellar ligament to tuberosity of tibia

■ Action
Extends leg

■ Nerve
Femoral nerve (L2–L4)

VASTUS MEDIALIS *(One of Quadriceps Femoris)*

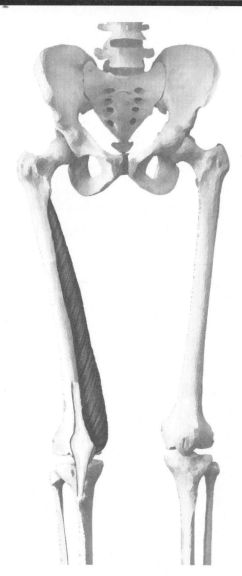
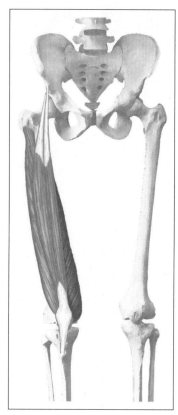

Hip, thigh, and leg—anterior view

■ Origin
Intertrochanteric line, medial lip of linea aspera of femur, medial intermuscular septum of adductor magnus and adductor longus, medial supracondylar ridge

■ Insertion
Medial border of the patella, then by patellar ligament into tibial tuberosity, medial condyle of tibia

■ Action
Extends leg, draws patella medially

■ Nerve
Femoral nerve (L2–L4)

Note: Although they are not anatomically separate,* the lower portion of the vastus medialis is referred to as the vastus medialis obliquus (VMO). Its action is to stabilize the patella and prevent its lateral dislocation.

*Noric, M. M., Mitchell, J., de Klerk, D. (1997) A comparison of the proximal and distal parts of the vastus medialis muscle. *Australian Journal of Physiotherapy*, 43(4):277–281.

VASTUS INTERMEDIUS *(One of Quadriceps Femoris)*

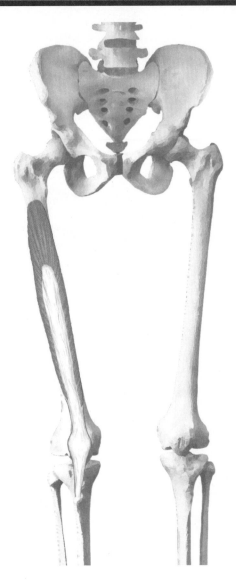

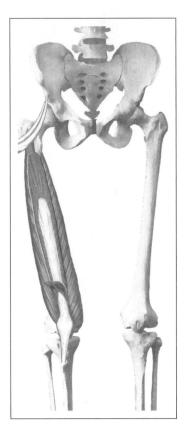

Quadriceps—rectus femoris cut

Hip, thigh, and leg—anterior view

■ Origin
Anterior and lateral surfaces of upper two-thirds of femur, lateral intermuscular septum, linea aspera, lateral supracondylar ridge

■ Insertion
Deep aspect of quadriceps tendon, then through patella to tibial tuberosity

■ Action
Extends leg

■ Nerve
Femoral nerve (L2–L4)

Note: A few bundles of fibers from this muscle insert onto the upper part of the joint capsule of the knee. They probably draw the capsule superiorly during extension of the leg, preventing it from binding in the joint. They are called *articularis genus*.

BICEPS FEMORIS *(Part of Hamstrings)*

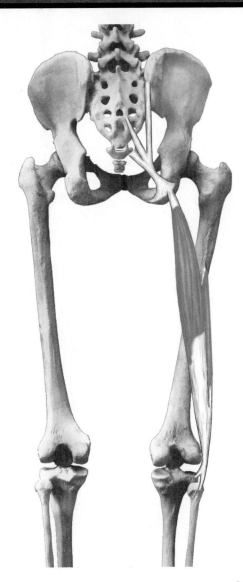

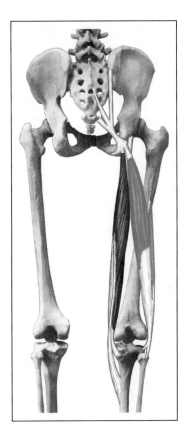

Posterior thigh muscles—hamstrings

Hip and thigh—posterior view

■ Origin
Long head—ischial tuberosity, sacrotuberous ligament

Short head—linea aspera, lateral supracondylar ridge, of femur lateral intermuscular septum

■ Insertion
Lateral side of head of fibula and lateral condyle of tibia

■ Action
Flexes leg, long head also extends thigh

■ Nerve
Long head—tibial part of sciatic nerve (S1–S3)

Short head—common peroneal part of sciatic nerve (L5, S1, S2)

Note: During walking or running, the hamstrings are used to slow down the leg at the end of its swing and prevent the trunk from flexing at the hip. They are susceptible to being strained by resisting the momentum of these body parts.

SEMITENDINOSUS *(Part of Hamstrings)*

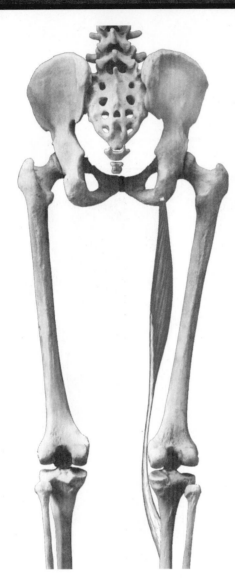

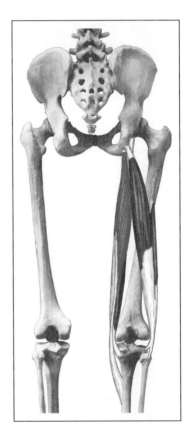

Posterior thigh muscles—hamstrings

Hip and thigh—posterior view

■ Origin
Ischial tuberosity

■ Insertion
Medial surface of shaft of tibia

■ Action
Flexes and slightly medially rotates leg; after flexion, extends thigh

■ Nerve
Tibial portion of sciatic nerve (L5, S1, S2)

See note on p. 181 and Relationship section on p. 176.

SEMIMEMBRANOSUS *(Part of Hamstrings)*

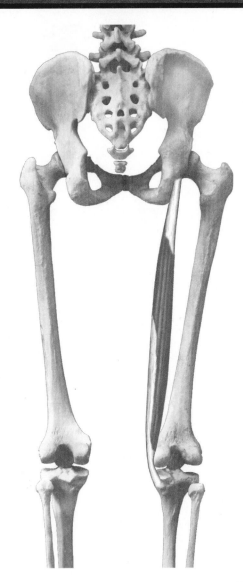

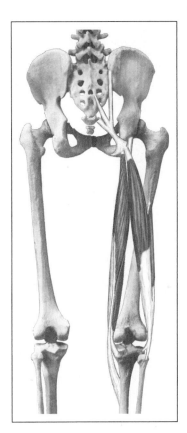

Posterior thigh muscles—hamstrings

Hip and thigh—posterior view

■ Origin
Ischial tuberosity

■ Insertion
Posterior part of medial condyle of tibia

■ Action
Flexes and slightly medially rotates leg; after flexion, extends thigh

■ Nerve
Tibial portion of sciatic nerve (L5, S1, S2)

See note on p. 181.

MUSCLES OF POSTERIOR THIGH

Hip and thigh—posterior view

1. Sciatic nerve
2. Quadratus femoris
3. Biceps femoris*
4. Semimembranosus*
5. Semitendinosus*
6. Tibial nerve
7. Common peroneal nerve

Note: The common peroneal nerve is exposed to compression and damage as it passes over the head of the fibula. The quadratus femoris, a lateral rotator, is included for reference.

*Hamstring muscles.

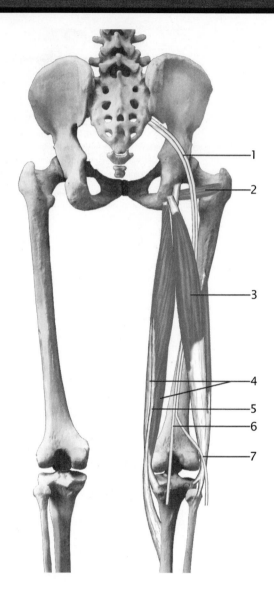

GRACILIS

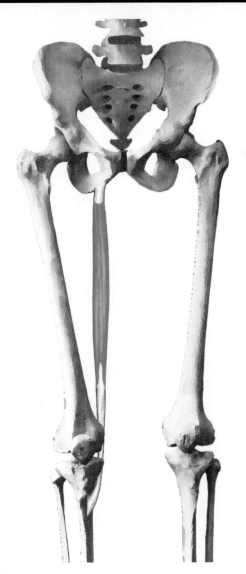

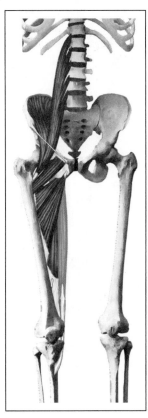

Hip flexors and adductors

Hip and thigh—anterior view

■ Origin
Lower margin of body and inferior ramus of pubis

■ Insertion
Upper part of medial surface of shaft of tibia

■ Action
Adducts thigh at hip joint and flexes leg, with leg flexed, assists in medial rotation

■ Nerve
Obturator nerve (L2–L4)

See Relationship section on p. 176.

PECTINEUS

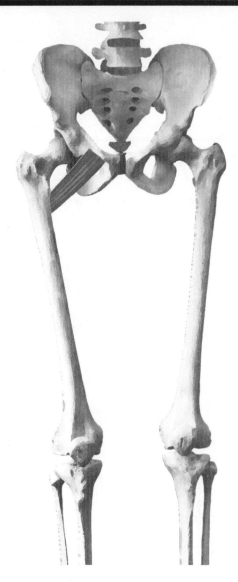

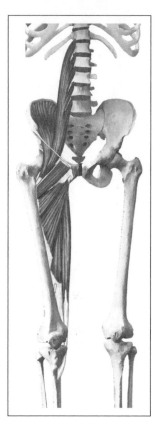

Hip flexors and adductors

Hip and thigh—anterior view

■ Origin
Pectineal line on superior ramus of pubis

■ Insertion
From lesser trochanter to linea aspera of femur

■ Action
Flexes thigh, assists in adduction when hip is flexed

■ Nerve
Femoral nerve (L2, L3), (sometimes a branch of obturator nerve)

Note: The rotational component of thigh muscle action depends upon the starting position of the hip joint. The pectineus, adductor longus, adductor brevis, and psoas major probably assist in medial rotation when the thigh is in anatomical position but may shift to assisting in lateral rotation as the thigh flexes and abducts.

The iliacus and adductor tubercle part of adductor magnus probably assist in medial rotation throughout the range of motion of the hip joint while the linea aspera part of the adductor magnus may be a slight lateral rotator.

ADDUCTOR LONGUS

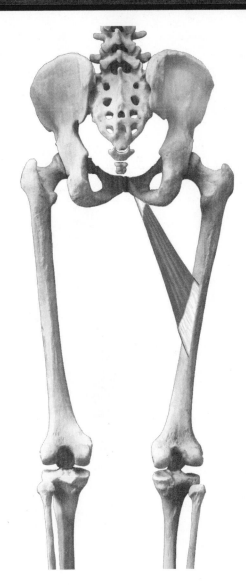

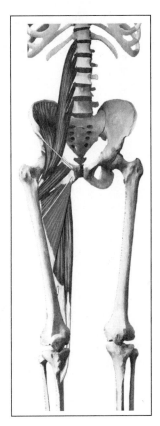

Hip flexors and adductors

Hip and thigh—posterior view

■ Origin
Anterior of body of pubis

■ Insertion
Medial lip of linea aspera

■ Action
Adducts, flexes thigh, assists in medial rotation*

■ Nerve
Obturator nerve (L2–L4)

*See note on p. 186.

ADDUCTOR BREVIS

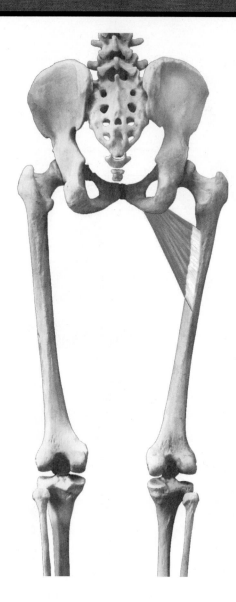

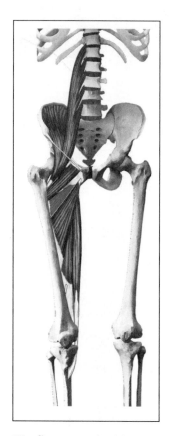

Hip flexors and adductors

Hip and thigh—posterior view

■ Origin
Outer surface of inferior ramus of pubis

■ Insertion
From below lesser trochanter to linea aspera and into proximal part of linea aspera

■ Action
Adducts thigh, assists in flexion, medial rotation*

■ Nerve
Obturator nerve (L2–L4)

*See note on p. 186.

ADDUCTOR MAGNUS

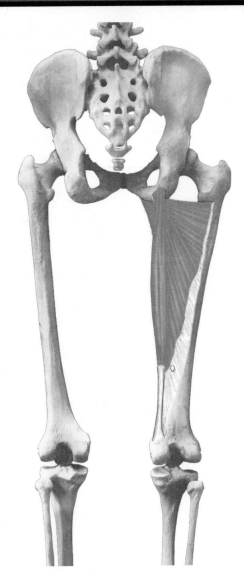

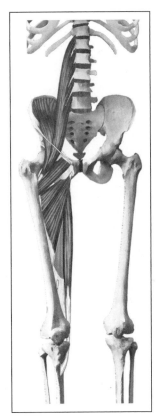

Hip flexors and adductors

Hip and thigh—posterior view

■ Origin
Inferior ramus of pubis, ramus and lower part of tuberosity of ischium

■ Insertion
Linea aspera, adductor tubercle of femur

■ Action
Adducts, extends thigh, lower portion (adductor tubercle insertion) assists in medial rotation*

■ Nerve
Obturator nerve (L2–L4), sciatic nerve

*See note on p. 186.

MUSCLES OF ANTERIOR THIGH

Anterior thigh

1. Psoas major
2. Iliacus
3. Tensor fascia latae
4. Inguinal ligament
5. Sartorius (cut)
6. Femoral artery, vein
7. Femoral nerve
8. Pectineus
9. Iliotibial band
10. Adductor brevis
11. Adductor longus
12. Adductor magnus
13. Gracilis
14. Rectus femoris (cut)
15. Vastus lateralis
16. Vastus intermedius
17. Vastus medialis
18. Patellar tendon

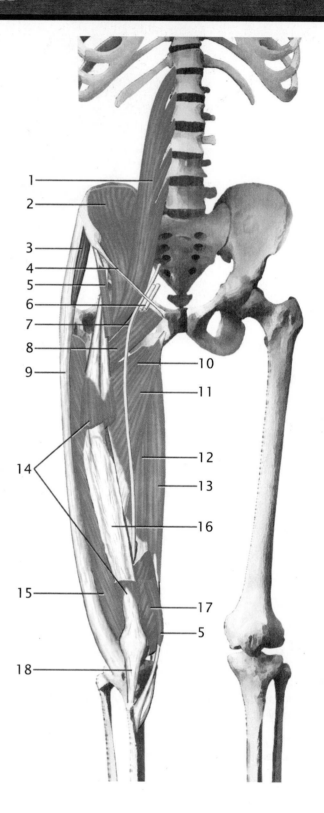

CHAPTER NINE

Muscles of the Leg and Foot

TIBIALIS ANTERIOR

Anterior and lateral leg muscles

Leg—anterolateral view

■ Origin
Lateral condyle of tibia, upper half of lateral surface of tibia, interosseous membrane

■ Insertion
Medial side and plantar surface of medial cuneiform bone, base of first metatarsal bone

■ Action
Dorsiflexes foot, inverts (supinates) foot

■ Nerve
Deep peroneal nerve (L4, L5, S1)

EXTENSOR HALLUCIS LONGUS

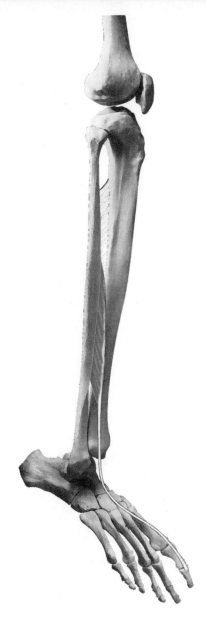

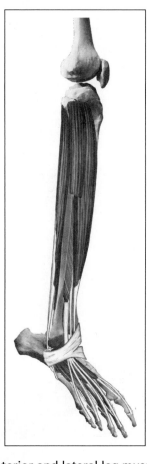

Anterior and lateral leg muscles
(extensor digitorum longus cut)

Leg—anterolateral view

■ Origin
Middle half of anterior surface of fibula and
interosseous membrane

■ Insertion
Base of distal phalanx of great toe

■ Action
Extends, hyperextends great toe, dorsiflexes and inverts
(supinates) foot

■ Nerve
Deep peroneal nerve (L4, L5, S1)

EXTENSOR DIGITORUM LONGUS

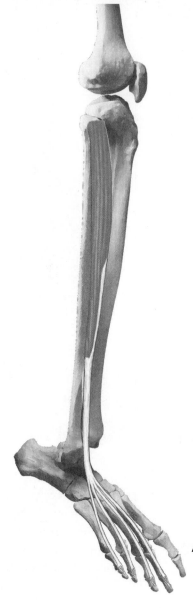

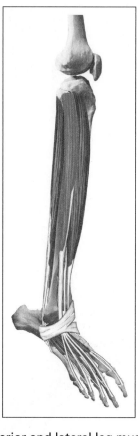

Anterior and lateral leg muscles

Leg—anterolateral view

■ Origin
Upper two-thirds of anterior surface of fibula, interosseous membrane, lateral condyle of tibia

■ Insertion
Along dorsal surface of four lateral toes, and then to bases of middle and distal phalanges

■ Action
Extends toes, dorsiflexes foot, everts foot

■ Nerve
Deep peroneal nerve (L4, L5, S1)

Note: The lower lateral part of this muscle makes a separate insertion onto the dorsal surface of the fifth metatarsal and is called *peroneus tertius*.

FIBULARIS TERTIUS *(Peroneus Tertius)*

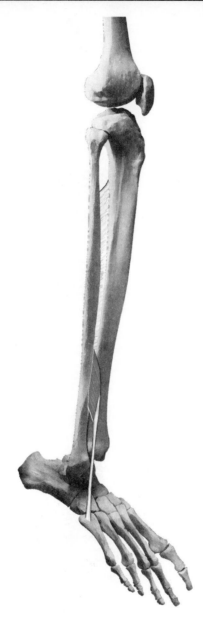

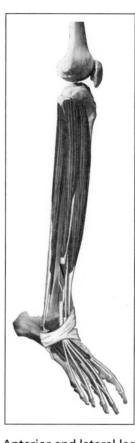

Anterior and lateral leg
muscles (extensor digitorum
longus tendon cut)

Leg—anterolateral view

■ Origin
Lower third of anterior surface of fibula and
interosseous membrane

■ Insertion
Dorsal surface of base of fifth metatarsal bone

■ Action
Dorsiflexes and everts foot

■ Nerve
Deep peroneal nerve (L4, L5, S1)

Note: This muscle is not present in all individuals. It may be
described as the fifth tendon of extensor digitorum longus.

GASTROCNEMIUS *(Part of Triceps Surae)*

Leg—posterior view

■ Origin

Lateral head—lateral condyle and posterior surface of femur

Medial head—popliteal surface of femur above medial condyle

■ Insertion

Posterior surface of the calcaneus

■ Action

Plantar flexes foot, flexes leg when foot is dorsiflexed and knee is extended

■ Nerve

Tibial nerve (S1, S2)

Superficial posterior
leg muscles

SOLEUS *(Part of Triceps Surae)*

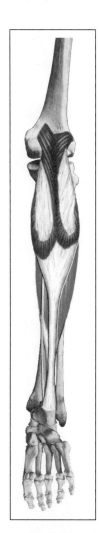

Superficial posterior
leg muscles

Leg—posterior view

■ Origin
Posterior surface of the tibia (soleal line), upper third of posterior surface of fibula, fibrous arch between tibia and fibula

■ Insertion
Posterior surface of the calcaneus

■ Action
Plantar flexes foot

■ Nerve
Tibial nerve (S1, S2)

PLANTARIS

Leg—posterior view

■ Origin
Lateral supracondylar ridge of femur, oblique popliteal ligament

■ Insertion
Posterior surface of the calcaneus

■ Action
Plantar flexes foot, flexes leg

■ Nerve
Tibial nerve (S1, S2)

Superficial posterior leg muscles (gastrocnemius cut and moved)

POPLITEUS

Deep posterior
leg muscles

Leg—posterior view

■ Origin
Lateral surface of lateral condyle of femur

■ Insertion
Upper part of posterior surface of tibia

■ Action
Rotates leg medially, flexes leg

■ Nerve
Tibial nerve (L4, L5, S1)

Note: Stern contends that this muscle stabilizes the knee by preventing lateral rotation of the tibia during medial rotation of the thigh while the foot is planted. Knee stabilization allows standing without undue fatigue to the quadriceps.

Reference: Stern, J. T. *Essentials of Gross Anatomy*, F. A. Davis Company, Philadelphia, 1988.

FLEXOR HALLUCIS LONGUS

Leg—posterior view

■ Origin
Lower two-thirds of posterior surface of shaft of fibula, posterior intermuscular septum, interosseous membrane

■ Insertion
Base of distal phalanx of great toe

■ Action
Flexes distal phalanx of great toe, assists in plantar flexing foot, inverts foot

■ Nerve
Tibial nerve (L5, S1, S2)

Note: This muscle is important in pushing off the surface in walking, running, jumping.

Deep posterior leg muscles

FLEXOR DIGITORUM LONGUS

Deep posterior leg muscles

Leg—posterior view

■ Origin
Medial part of posterior surface of tibia

■ Insertion
Bases of distal phalanges of second, third, fourth, and fifth toes

■ Action
Flexes distal phalanges of lateral four toes, assists in plantar flexing foot, inverts foot

■ Nerve
Tibial nerve (L5, S1)

TIBIALIS POSTERIOR

Leg—posterior view

■ Origin
Lateral part of posterior surface of tibia, interosseous membrane, proximal half of posterior surface of fibula

■ Insertion
Tuberosity of navicular bone, cuboid, cuneiforms, second, third, and fourth metatarsals, sustentaculum tali of calcaneus

■ Action
Plantar flexes, inverts foot

■ Nerve
Tibial nerve (L5, S1)

Deep posterior leg muscles

FIBULARIS LONGUS *(Peroneus Longus)*

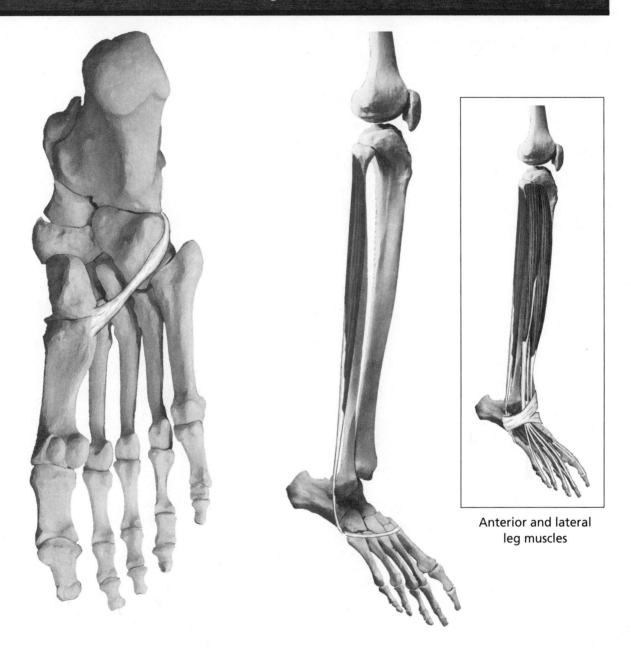

Anterior and lateral
leg muscles

Foot—plantar view

■ Origin
Upper two-thirds of lateral surface of fibula

■ Insertion
Lateral side of medial cuneiform, base of first metatarsal

■ Action
Plantar flexes, everts foot

■ Nerve
Superficial peroneal nerve (L4, L5, S1)

FIBULARIS BREVIS *(Peroneus Brevis)*

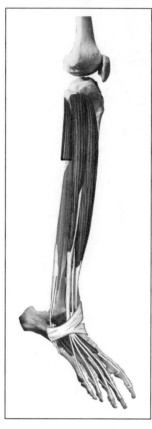

Anterior and lateral leg muscles
(Fibularis longus cut)

Leg—anterolateral view

■ **Origin**
Lower two-thirds of lateral surface of fibula

■ **Insertion**
Lateral side of base of fifth metatarsal bone

■ **Action**
Everts, plantar flexes foot

■ **Nerve**
Superficial peroneal nerve (L5, S1, S2)

EXTENSOR DIGITORUM BREVIS

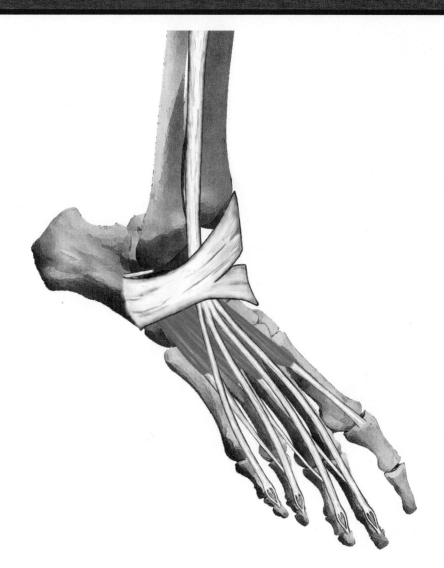

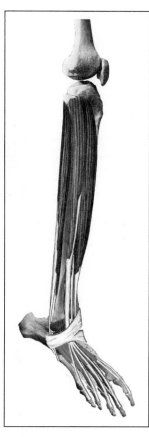

Anterior and lateral leg muscles

Foot—anterolateral view

■ Origin
Anterior and lateral surfaces of calcaneus, lateral talocalcaneal ligament, inferior extensor retinaculum

■ Insertion
Into base of proximal phalanx of great toe; into lateral sides of tendons of extensor digitorum longus of second, third, and fourth toes

■ Action
Extends the four toes

■ Nerve
Deep peroneal nerve (S1, S2)

ABDUCTOR HALLUCIS *(First Layer)*

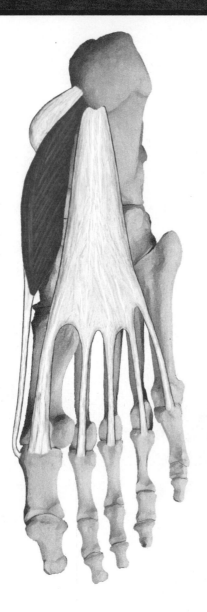

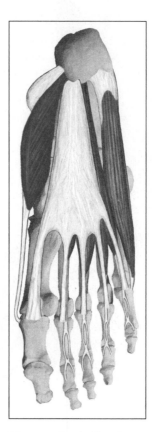

Plantar view—first layer

Foot—plantar view

■ Origin
Tuberosity of calcaneus, flexor retinaculum, plantar aponeurosis

■ Insertion
Medial side of base of proximal phalanx of great toe

■ Action
Stabilizes great toe (with adductor hallucis)

■ Nerve
Medial plantar nerve (L4, L5)

Note: The muscles of the sole of the foot can be divided into four layers (from superficial to deep):

First layer—abductor hallucis, flexor digitorum brevis, abductor digiti minimi

Second layer—quadratus plantae, lumbricales (tendons of flexor hallucis longus and flexor digitorum longus pass through this layer)

Third layer—flexor hallucis brevis, adductor hallucis, flexor digiti minimi brevis

Fourth layer—interossei (tendons of tibialis posterior and fibularis longus pass through this layer)

FLEXOR DIGITORUM BREVIS *(First Layer)*

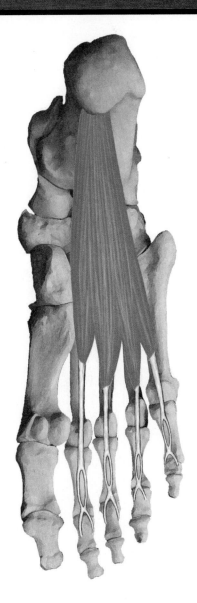

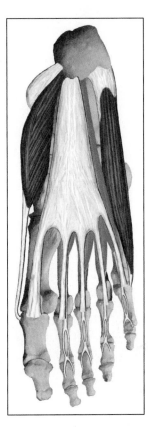

Plantar view—first layer

Foot—plantar view

■ Origin
Tuberosity of calcaneus, plantar aponeurosis

■ Insertion
Sides of middle phalanges of second to fifth toes

■ Action
Flexes proximal phalanges and extends distal phalanges of second through fifth toes

■ Nerve
Medial plantar nerve (L4, L5)

ABDUCTOR DIGITI MINIMI *(First Layer)*

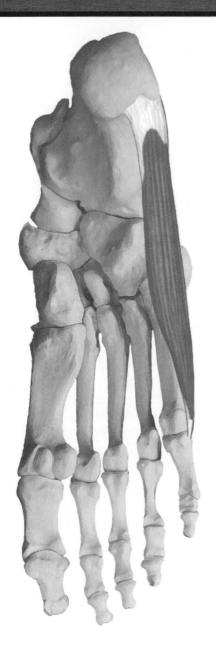

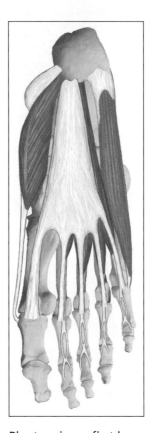

Plantar view—first layer

Foot—plantar view

■ Origin
Tuberosity of calcaneus, plantar aponeurosis

■ Insertion
Lateral side of proximal phalanx of fifth toe

■ Action
Abducts fifth toe

■ Nerve
Lateral plantar nerve (S1, S2)

QUADRATUS PLANTAE *(Second Layer)*

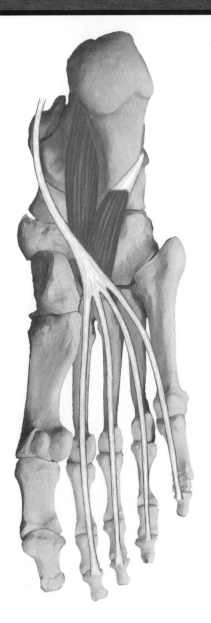

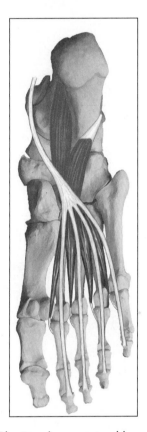

Plantar view—second layer

Foot—plantar view

■ Origin
Medial head—medial surface of calcaneus

Lateral head—lateral border of inferior surface of calcaneus

■ Insertion
Lateral margin of tendon of flexor digitorum longus

■ Action
Flexes terminal phalanges of second through fifth toes

■ Nerve
Lateral plantar nerve (S1, S2)

LUMBRICALS *(Second Layer)*

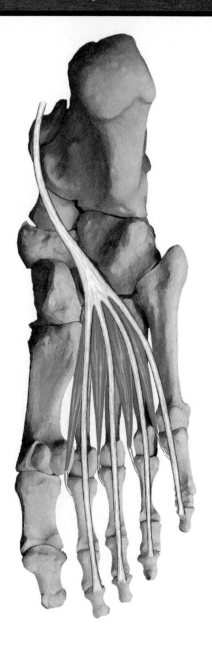

Plantar view—second layer

Foot—plantar view

■ Origin
Tendons of flexor digitorum longus

■ Insertion
Dorsal surfaces of proximal phalanges

■ Action
Flex proximal phalanges of second through fifth toes

■ Nerve
First lumbricalis—medial plantar nerve (L4, L5)

Second through fifth lumbricales—lateral plantar nerve (S1, S2)

FLEXOR HALLUCIS BREVIS *(Third Layer)*

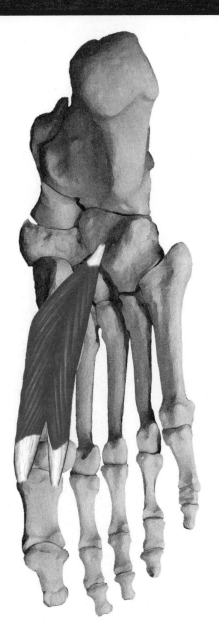

Plantar view—third layer

Foot—plantar view

■ Origin
Cuboid bone, lateral cuneiform bone

■ Insertion
Medial part—medial side of base of proximal phalanx of great toe

Lateral part—lateral side of base of proximal phalanx of great toe

■ Action
Flexes proximal phalanx of great toe

■ Nerve
Medial plantar nerve (L4, L5, S1)

Note: The tendons of insertion contain sesamoid bones.

ADDUCTOR HALLUCIS *(Third Layer)*

Plantar view—third layer

Foot—plantar view

■ Origin
Oblique head—second, third, and fourth metatarsal bones, and sheath of fibularis longus tendon

Transverse head—plantar metatarsophalangeal ligaments of third, fourth, and fifth toes, and transverse metatarsal ligaments

■ Insertion
Lateral side of base of proximal phalanx of great toe

■ Action
Stabilizes great toe (with abductor hallucis)

■ Nerve
Lateral plantar nerve (S1, S2)

FLEXOR DIGITI MINIMI BREVIS *(Third Layer)*

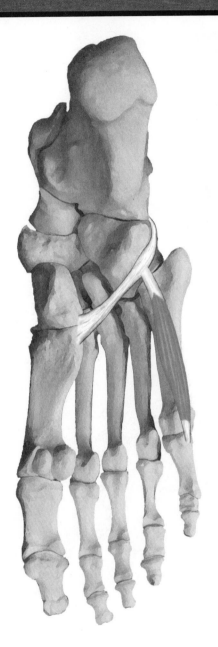

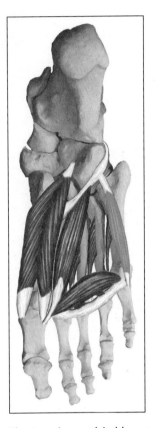

Plantar view—third layer

Foot—plantar view

■ Origin
Base of fifth metatarsal, sheath of fibularis longus tendon

■ Insertion
Lateral side of base of proximal phalanx of fifth toe

■ Action
Flexes proximal phalanx of fifth toe

■ Nerve
Lateral plantar nerve (S1, S2)

DORSAL INTEROSSEI *(Fourth Layer; Four Muscles)*

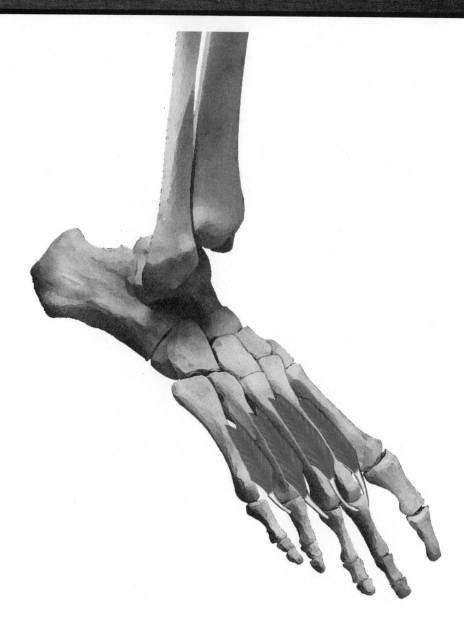

Foot—anterolateral view

■ **Origin**

Adjacent sides of metatarsal bones

■ **Insertion**

Bases of proximal phalanges

First—medial side of proximal phalanx of second toe

Second, third, fourth—lateral sides of proximal phalanges of second, third, and fourth toes

■ **Action**

Abduct toes, flex proximal phalanges

■ **Nerve**

Lateral plantar nerve (S1, S2)

PLANTAR INTEROSSEI *(Fourth Layer; Three Muscles)*

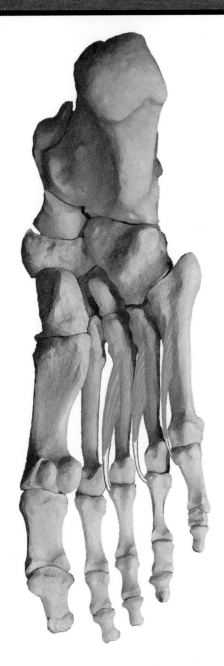

Foot—plantar view

■ Origin
Bases and medial sides of third, fourth, and fifth metatarsal bones

■ Insertion
Medial sides of bases of proximal phalanges of same toes

■ Action
Adduct toes, flex proximal phalanges

■ Nerve
Lateral plantar nerve (S1, S2)

Alphabetical Listing of Muscles

Index

Notes

Notes

Notes

Notes